BALTIMORE
1991

TECHNICAL PAPERS

1991 ACSM-ASPRS ANNUAL CONVENTION

Volume 1
Surveying

ACSM 51st Annual Convention
ASPRS 56th Annual Convention

Copyright © 1991 by the American Congress on Surveying and Mapping and the American Society for Photogrammetry and Remote Sensing. All rights reserved. Reproductions of this volume or any parts thereof (excluding short quotations for use in the preparation of reviews and technical and scientific papers) may be made only after obtaining the specific approval of the publishers. The publishers are not responsible for any opinions or statements made in the technical papers.

Permission to Photocopy: The copyright owners hereby give consent that copies of this book, or parts thereof, may be made for personal or internal use, or for the personal or internal use of specific clients. This consent is given on the condition, however, that the copier pay the stated copy fee of $2 for each copy, plus 10 cents per page copied (prices subject to change without notice), through the Copyright Clearance Center, Inc., 27 Congress St., Salem, MA 01970, for copying beyond that permitted by Sections 107 or 108 of the U.S. Copyright Law. This consent does not exceed to other kinds of copying, such as copying for general distribution, for advertising or promotional purposes, for creating new collective works, or for resale.

When reporting copies from this volume to the Copyright Clearance Center, Inc., please refer to the following code: ISBN 0-944426-39-5/91/2 + .10.

ISBN 0-944426-38-7
ISBN 0-944426-39-5

Published by
American Congress on Surveying and Mapping
and
American Society for Photogrammetry and Remote Sensing
5410 Grosvenor Lane
Bethesda, MD 20814-2160
USA
Printed in the United States of America

TABLE OF CONTENTS

	PAGE
Anderson, Kenneth W. Are There Really Original Bearing Trees in Southern Indiana?	1
Balazs, E. I. Test Results of the NA2000 Leveling System	6
Ball Jr., William E. Design Criteria for Three-Dimensional Coordinate Computational Software for Future Land Information Systems	13
Barnes, Grenville A Comparative Evaluation Framework for Cadastral Systems in the U.S. (Abstract)	24
Brinker, Dexter M. Spherical Trigonometry - Key to Understanding World Geography, Navigation and Control Surveys	25
Brown, John M.; Bethel, James S. Blunder Detection in a Small Geodetic Network	31
Burkholder, Earl F. Design of a Local Coordinate System for Surveying, Engineering and LIS/GIS (Abstract)	37
Chong, Albert K. Accurate Height Transfer from Bench Marks to Awkward Sites Using GPS	38
Cole, George M.; Speed, F. Michael Use of Constituent Analysis for Estimation of Tidal Data by Simultaneous Short-Term Observations	44
Crossfield, James K.; Parks, Wesley; Littell, Barbara S. GPS Project Design: The Owens Valley Experience	54
Doytsher, Y.; Shmutter, B. Incorporating Known External Data in Information Digitized from Cadastral Maps	61
Dracup, Joseph F. Some Notes on Adjustments, Weights, and Accuracies	69

Ferguson, C. Roger; Caron, Phillip B. 81
Tax Mapping on the Connecticut Coordinate System
With Scale Improvements, Using Approximately Scaled
Air Photographs, Surveyors Maps, and
Quadrangle Sheets

Gary, George E. 90
"WASHMORE" - Washington Monument Resurvey Expedition

Glover, Charles C. 96
Leveling Without a Level Instrument (A Trigonometric
Method) (Abstract)

Greenfeld, Joshua S. 97
Developing Transverse Mercator Projection Tables
for NAD'83

Hamilton, John 107
Geodetic Surveying in Southwestern Pennsylvania

Hartzheim, Paul J.; Hothem, Larry D.; Kyle, Dale W. 121
Results of Wisconsin High Precision Geodetic
Network Survey (Abstract)

Hattori, Susumu; Hasegawa, Hiroyuki; Uesugi, Kouhei 122
A New Plane Table System - CG Plane Table

Jeyapalan, K.; Erck, E. S. 130
Elevation Difference Recovery Tests from GPS
Observation and Gravimetry

Kaufman, Donald J. 144
The Receivership Solution: Protecting the Rights
of Minority Stockholders

Krupnik, A.; Shmutter, B. 151
Automated Enhancement of Shapes of Registration
Parcels

Leick, Alfred; Liu, Quanjiang; Burkholder, Earl 157
Systematic Effects of Single-Frequency GPS in the
Continuous Kinematic Mode (Abstract)

Livermore, Marlin 158
The Public Domain Land Tenure System in the
United States

Loy, Jim 166
Total Quality Development in a Land Survey Firm

Marth Sr., Richard B. 173
Topographic Surveyor - Transitioning to the 90s

Martin, Douglas M.; Speed, F. Michael; Thurlow, Carroll I. 179
Tidal Datum Computation in Low Tide Range Areas (Abstract)

McReynolds, Jon E. 180
Open-End Electronic Survey Traversing

Moreno, Ricardo J. 187
A Formal Model for Land Titles and Interests in Land (Abstract)

Moussa, Osama M.; Attabi, Mohamed M. H.; Soliman, Ehab H. 188
Stability Analysis of Structures Over and Around the Egyptian Subway

Niles, Anthony 191
Dynamic Positioning Systems Photogrammetric Test Course

Onsrud, Harlan J. 197
Evidentiary Admissibility and Reliability of Products Generated from GIS

Paiva, J. V. R. 202
An Intelligent Solution for Kinematic GPS Mapping

Parks, W. W.; Crossfield, J. K.; Littell, B. S. 210
The Owens Valley GPS Control Survey (Abstract)

Rodine, Corwyn J.; Wahl, Jerry L.; Parker, Blair; Hintz, Raymond J.; Blanchard, Barry M. 211
Progress Report on the Development of an Integrated PLSS Cadastral Measurement Management and Retracement Survey Software System

Roe, Gene V. 219
A Grass Roots Strategy for Establishing the Surveying Professional's Proper Role Vis-a-Vis LIS

Scruggs, Robert 225
Integrating GPS Into the Bureau of Land Management's Cadastral Survey Program

Shackelford, Michael G. 231
1791 District of Columbia Boundary Survey

Shmutter, B.; Doytsher, Y. 241
A New Method for Matching Digitized Cadastral Maps

Shreves, Dennis D. 247
Back from the Brink - Ferret Habitat Survey
at Wind Cave National Park

von Meyer, Nancy 255
A Discussion of Dualism in Public Sector Land
Information

Wahl, Jerry L.; Hintz, Raymond J.; Rodine, Corwyn J. 259
Development of an Electronic Field Book for
Cadastral Retracement Surveys

Wijayratne, Indrajith 269
Conversion of NAD 27 State Plane Coordinates to
NAD 83 State Plane Coordinates by a 2-D Projective
Transformation

Winslett, Cari; Pearsall, Robert J. 277
Geodetic Control: Traditional Technology
Transformed for Today

Yates, Jeffrey F. 280
The 1989 Mt. McKinley, Alaska, GPS Expedition

Zilkoski, David B. 290
A Priori Estimates of Standard Errors of
Leveling Data

AUTHOR INDEX

Author	Page
Anderson, Kenneth W.	1
Attabi, Mohamed M. H.	188
Balazs, E. I.	6
Ball Jr., William E.	13
Barnes, Grenville	24
Bethel, James S.	31
Blanchard, Barry M.	211
Brinker, Dexter M.	25
Brown, John M.	31
Burkholder, Earl F.	37, 157
Caron, Phillip B.	81
Chong, Albert K.	38
Cole, George M.	44
Crossfield, James K.	54, 210
Doytsher, Y.	61, 241
Dracup, Joseph F.	69
Erck, E. S.	130
Ferguson, C. Roger	81
Gary, George E.	90
Glover, Charles C.	96
Greenfeld, Joshua S.	97
Hamilton, John	107
Hartzheim, Paul J.	121
Hasegawa, Hiroyuki	122
Hattori, Susumu	122
Hintz, Raymond J.	211, 259
Hothem, Larry D.	121
Jeyapalan, K.	130
Kaufman, Donald J.	144
Krupnik, A.;	151
Kyle, Dale W.	121
Leick, Alfred	157
Littell, Barbara S.	54, 210
Liu, Quanjiang	157
Livermore, Marlin	158
Loy, Jim	166
Marth Sr., Richard B.	173
Martin, Douglas M.	179
McReynolds, Jon E.	180
Moreno, Ricardo J.	187
Moussa, Osama M.	188
Niles, Anthony	191
Onsrud, Harlan J.	197
Paiva, J.V. R.	202
Parker, Blair	211
Parks, W. W.	54, 210
Pearsall, Robert J.	277
Rodine, Corwyn J.	211, 259
Roe, Gene V.	219
Scruggs, Robert	225
Shackelford, Michael G.	231
Shmutter, B.	61, 151, 241

Shreves, Dennis D.		247
Soliman, Ehab H.		188
Speed, F. Michael	44,	179
Thurlow, Carroll I.		179
Uesugi, Kouhei		122
von Meyer, Nancy		255
Wahl, Jerry L.	211,	259
Wijayratne, Indrajith		269
Winslett, Cari		277
Yates, Jeffrey F.		280
Zilkoski, David B.		290

ARE THERE REALLY ORIGINAL BEARING TREES IN SOUTHERN INDIANA?

Kenneth W. Anderson, PLS
USDA-Forest Service
RR#11   Box 782
Bedford, Indiana 47421

BIOGRAPHICAL SKETCH

Kenneth W. Anderson, a registered Land Surveyor in Indiana, has been the forest Land Surveyor on the Wayne/Hoosier National Forest since 1978. He received a B.S. degree in Forest Management from the University of Minnesota in 1960. The past thirty years he has been employed by the Forest Service in several positions. His interest in the original survey system ultimately led him to his present position and his search for original bearing trees.

ABSTRACT

Existing Bearing Trees from the original government surveys in southern Indiana (1800-1820) are few in number, but while less than one percent of these trees still exist, these few show that the survey system has survived intact for almost two centuries. To find and prove an original bearing tree is not an easy task. It involves hours of records search as well as close inspection of the tree itself, and the assumption might be that only the largest and strongest trees would have survived for 180 odd years. But the truth is that the less dominate trees are often the survivors.

Searching for and finding original bearing trees is an exciting project that stirs the hearts and minds of all too few surveyors these days. In our present world of electronic wizardry, the "BT"s that made much of this possible are seldom considered.

INTRODUCTION

Nearly all land surveyors are aware that the government surveyors in the Public Land States established bearing trees as a matter of course. These legendary trees, selected between 1790 and 1900, have mostly died and rotted away so that today's surveyor can work a lifetime without ever actually seeing one. This is unfortunate because they were part of our basic survey system.

The original government surveys in southern Indiana were done in the 1800-1820 era. At each section and one-quarter corner a wood post was set and two bearing trees were established. It is assumed that the trees were "marked" by the original surveyors, however no definite evidence has been found to that effect. I believe they were "marked" somehow, perhaps with blazes, hacks, notches, etc., but probably not scribed as was the practice later on.

Since arriving in southern Indiana, I have been looking for "original bearing trees." Having seen many scribed bearing trees in Minnesota and Wisconsin in previous years, I looked for a similar type of evidence in Indiana, but this proved to be faulty reasoning. I soon learned there are few of these trees left, and when found, there is difficulty in proving they are in fact the ones called for. It is a different process and there are always some questions left unanswered.

To date I have found evidence of three trees which I feel confident are "original bearing trees" from the government surveys of the early 1800s. In addition there are three others which may be original bearing trees.

The proof is "circumstantial" at best. In other words, there are no distinguishing marks or scribes on the trees to say that they are "the tree." Proof is gained by research of records over the past 180 years and supplemented by measurements to other corner positions. Though not without error, it is the best evidence available, and the ones that have been found have supported the basic survey system as originally established.

Surprisingly, of the six trees found, five are American Beeches, and only one is a Black Oak. Who would have thought that the Beech trees would outlast the more durable Oaks, Gums, and Hickories? The reason, I believe, is that the other species were in demand for other uses--lumber, posts, firewood, whereas the Beech was left alone to live a long slow life.

A CASE HISTORY

The one-quarter corner for Sections 26 and 35 T1N R1W Orange County, approximately two miles from the Indiana Initial Point, was established in 1805 by Ebenezer Buckingham, government surveyor, who set a wood post at average distance and selected two bearing trees, a Beech 20 inches in diameter S80W 15 links and a Beech 12 inches in diameter S60E 43 links.

In 1868, John Frazer, County Surveyor, set a stone at the corner point from the existing original bearing trees. Fifteen years later in 1883 C. H. Pinnick found the Beech bearing tree 30 inches in diameter S60E 43 links and established a new bearing tree Chestnut 24 inches S65W 31 links.

In 1888 L. B. White recovered the "Original Beech" bearing tree.

After the county surveyor's work in the late eighteen hundreds no written record of the corner was found. A survey in 1981 by a private surveyor under contract to the Forest Service recovered a corner stone and a large old 40 inch plus Beech tree. The stone was unmarked but was authentic and at record distance from the tree. Analysis of the situation indicated that the Beech tree was, in fact, an original bearing tree from the 1805 government survey by Ebenezer Buckingham.

This tree is on privately owned land and within two hundred feet of a rural home site with pasture lands on three sides. Why it survived in this spot for over one hundred years with obvious occupation use is a mystery. In 1981 it was alive but ailing and since then has died and now is nothing more than a dead snag.

A SECOND BEECH BEARING TREE

The second case history is about another Beech tree and is only four miles northeasterly of the first. This one is again at a one-quarter corner on the east side of Section 12, T1N, R1W, Orange County, Indiana on the Meridian Line four and one-half miles north of Indiana's Initial Point.

While the first tree was "exposed to the world" for nearly all the 185 years, this tree, in contrast, has been protected or semi-protected most of the same time period. It is on the southerly edge of Pioneer Mothers Memorial Forest about two miles southeasterly of Paoli, Indiana. This is a relatively undisturbed tract of land that has been in the Hoosier National Forest under administration of the Forest Service since 1941. Prior to that time, the land was part of a Cox family estate for seventy years or more, and they kept a portion of their lands undisturbed in a natural state.

The original government survey was done in 1806 by Arthur Henrie who set a post at 40.00 chains and established two bearing trees, a Hackberry 20 inches in diameter bears S20E 17 links and a Beech 10 inches diameter bears N37W 5 links. In 1830 a deed for adjacent lands called for "beginning at the half-mile stake on the Meridian line in Section Twelve." The next transfer of the land was in 1857 and calls for "beginning at the half-mile corner on the east side of said Section Twelve on the Meridian at a stone." This description, or one very similar, was used several times in the land deeds prior to 1941. The corner stone monument did, in fact, exist until some time in the 1930-1940 era.

Evidence of the stone and bearing trees is called for in Orange County Survey records by a 1863 survey done by J.H. Lindsey, County Surveyor who noted "the old trees" and a corner 40 rods to the west. After that no record of the corner was found until 1935 when a surveyor/compassman for the Forest Service found a twenty-four inch Beech but made no mention of the corner stone. He may not have been aware that the tree he was looking at was an original bearing tree.

In the 1940s the State of Indiana and the US Government acquired lands southwesterly and northwesterly respectively. In neither case was there any definite mention of the corner stone monument or the Beech tree. At some point, date unknown, the State of Indiana set a Right of Way 4"x4" concrete monument S70E 3.8 feet from the Beech tree. It is unknown whether it was set to replace the corner stone or was set at the juncture of the obvious occupation lines. There are several stones by the tree and ROW marker, but none could be found with marks on it.

A 1989 survey of the Pioneer Mothers Memorial Forest recovered several corner monuments, two of which were used to prove the location of the Beech tree. The corner stone forty rods (660 feet) west of the one quarter corner was found at 659.13 feet S89 degrees 33 minutes W from the corner point by the Beech tree. In addition the southeast section corner stone for section twelve was recovered about four inches underground and the measured distance was 2633.46 feet.

The surveying data was sufficient to place the Beech tree within the circle of the area to be searched. When no other evidence was found, it was decided that the Beech tree was an original bearing tree, and that the corner monument should be set at record distance from it. As in the first case, the tree is dead and now consists of a 28 inch diameter snag about fifteen feet tall.

THE THIRD HISTORY

The third original bearing tree was found about two years ago and is alive and healthy at this date. It was found by an elderly surveyor on one of his jobs in Martin County, Indiana. The tree, a Black Oak, is on a hilltop and on the township line for T3N R3W and was established in December 1804 by Levi Barber, government surveyor. He called for a post corner and bearing trees of Hickory 13 inches in diameter S48W 3 links and Black Oak 20 inches in diameter N8E 14 links. This corner was set 60 links east of the section corner for sections 34 and 35 T4N R3W. There is no known county survey data or other survey information for this particular corner.

Evidence to support this tree is limited but seems reasonable. There is the tree, forty inches plus in diameter and a very dominant feature at the fence intersection from east, west and south. Second, there is some limited survey data available to position the corner in that area. Third, a second old fence intersection is at record distance of sixty links to the west. Lastly, study of the aerial photos indicates that it is within the realm of possibility. Unfortunately no corner monuments were found at either corner to definitely prove the positions.

This is limited evidence to go on so it becomes a judgement call, but until some other evidence definitely proves or disproves the tree, I will consider it one of the original bearing trees.

OTHER POSSIBLE TREES

There are other trees that may be original bearing trees. At this time the evidence is too limited to definitely decide one way or another. In addition to existing physical trees or snags, there are stumps and stumpholes that are recovered during surveying activities.

CONCLUSION

Bearing trees from the original government surveys in the early eighteen hundreds still do exist in a few places. As you would expect, these two hundred-year plus trees are not easily found. When they are found, the evidence to prove that they are in fact original bearing trees is not easy to come by. It becomes a judgement call based on circumstantial evidence recovered as well as the size, shape, and character of the tree itself.

The fact that any exist after one hundred seventy some odd years of farming, logging, road building and natural weather causes, such as tornadoes, is in itself quite remarkable. A more intensive search may well reveal others that are unknown at this time.

Because there are so few available and because much survey work is in urban areas, many land surveyors in eastern states never see an original bearing tree, or if they do, they are unaware of what they are seeing.

My recommendation to the surveyors of today is to go and seek out an original bearing tree. It could prove to be of real interest to you.

TEST RESULTS OF THE NA2000 LEVELING SYSTEM

E.I. Balazs
National Geodetic Survey, NOS, NOAA
Rockville, MD 20852

ABSTRACT

Wild Leitz has developed a leveling system that measures, calculates, and records electronically. The system has been tested over a well-established network of bench marks where height differences have previously been determined several times by different precise leveling instruments. The new results indicate that the system could be used for precise leveling surveys if the leveling rods were of one-piece construction and the material of the rods had a lower coefficient of thermal expansion.

INTRODUCTION

In January 1990 Wild Leitz USA, Inc. requested that the Federal Geodetic Control Committee (FGCC) test the newly developed electronic digital leveling instrument, the NA2000. The test was performed during the week of June 25-29, 1990.

This unique instrument reads special bar codes on its leveling rod to provide a totally automatic digital readout of height. The instrument is a self-leveling level with a built-in computer for recording the observations. The observations are under program control. Rod readings are automatically repeated and displayed until the observer pushes the record button or the required number is reached. The leveling rods are made of fiberglass, total length of the three sections is 4.05 m.

STATIC TEST

The instrument and one rod were set up in a special, climate-controlled room (known as the "tape tunnel") in the Metrology Building at the National Institute of Standards and Technology (NIST) in Gaithersburg, Maryland. The following persons were in attendance: William Hollinshead and Gary Affolter, Wild Leitz USA, Inc.; Charles J. Fronczek, NIST; Harold G. Beard, Charles G. Glover, Orland Murray, and Jerry L. Pryor, NGS Operations Branch; and David B. Zilkoski, Rodney J. Lee, and I, NGS Vertical Network Branch.

Results from the static test are being evaluated by Messrs. Zilkoski and Fronczek. A final report on the static test will be published soon.

REPEATABILITY TEST

Two permanent bench marks, 71 m apart, of the Vertical Test Network on the NIST grounds were selected for this test. One bench mark, NRC 3, is a steel rod, driven to a depth of 25 ft and protected by a 1 inch diameter steel pipe. The other, RV 3, is a rivet cemented in the headwall of a culvert. The height difference between the two bench marks can be measured with one setup.

One NA2000 instrument and two rods were used. The rods are 4.05 m high, having three linkable units, each 1.35 m. Five observers measured the height difference between the two bench marks. Four observers made four setups each, one made five setups. Two setups for each observer were balanced sight lengths, two were unbalanced. The unbalanced setups differed by 5-6 m. The mean height differences and the standard deviations were:

1. From 10 balanced setups: 2.0332 m +/- 0.30 mm
2. From 11 unbalanced setups: 2.0329 m +/- 0.60 mm
3. From 21 setups: 2.0331 m +/- 0.52 mm
4. From 10 setups measured with
   4 different levels in 1973-74:   2.0310 m +/- 0.22 mm
5. From 2 setups with Ni002   2.0320 m

Twenty of the 21 measurements agreed within 1 mm or less, which is the requirement for first-order, class II surveys of a one-setup section. One unbalanced measurement differed by 1.6 mm from the mean of all measurements.

The reason for the excellent agreement between balanced and unbalanced measurements is that the correction for collimation error was applied to each rod reading by the instrument before the reading was recorded. Corrections for other systematic errors were not applied to the NA2000 measurements.

LOOP CLOSURE TEST

Five loops of the NIST test network (Figure 1) were releveled with two NA2000 instruments, run simultaneously, reading the same two rods. All sections were leveled in both forward and backward directions with both instruments. The observers and rod persons walked between setups during these test leveling. The loop misclosures, the allowable tolerances for first-order, class I surveys in mm, and the length of the loop in km are:

Loop	Serial No. 85394		Serial No. 85399	
1	+2.1 (4.0)	0.99	+2.4 (4.0)	0.98
2	-1.7 (3.8)	0.93	0.0 (3.8)	0.93
3	+1.2 (3.1)	0.62	+1.8 (3.1)	0.62
4	+0.3 (4.2)	1.09	+0.1 (4.2)	1.09
5	+0.9 (4.4)	1.22	-1.7 (4.4)	1.22

The standard deviations of a 1 km double-run section, computed from forward-backward measurements of 15 sections, are: 1.23 mm for instrument 85394 and 0.95 mm for 85399.

Seven of the eleven bench marks leveled by the two NA2000 instruments were leveled in 1968, 1969, 1970, 1973, and 1974. Corrections for systematic errors were applied to all measurements, but because the NA2000 systems rods were not calibrated, rod and temperature corrections were not computed. Bench mark F 113, a disk set in bedrock, was assumed stable, the two levelings of the five loop net were adjusted independently to determine two sets of new heights. The height differences in millimeters between old and new (NA2000) heights are tabulated below for both instruments.

Serial No. 85394

Bench Mark	1968	1969	1970	1973	1974
F 113	0.00	0.00	0.00	0.00	0.00
NBS 102	-0.02	-0.06	-0.15	-0.07	0.64
L 113	-0.63	0.25	-0.61	-0.55	-2.25
NBS 101	-0.12	-0.15	-0.51	0.60	-1.79
RV 2	15.13	14.91	13.03	9.11	7.20
RV 3	-3.61	-1.30	-2.03	0.89	-0.70
NRC 3	-0.23	-0.50	1.18	0.38	-0.82

Serial No. 85399

Bench Mark	1968	1969	1970	1973	1974
F 113	0.00	0.00	0.00	0.00	0.00
NBS 102	1.01	1.14	1.05	1.14	2.84
L 113	1.13	2.00	1.15	1.21	-0.49
NBS 101	1.80	1.44	2.55	2.50	0.16
RV 2	18.13	17.91	16.03	12.11	10.20
RV 3	-2.72	-0.40	-1.14	1.79	0.20
NRC 3	0.02	-0.25	1.43	0.62	-0.58

Except for RV 2, which was damaged by lawn mowers and will be replaced, all bench marks indicate stability between levelings. The heights determined by the NA2000 systems indicate the same stability as seen in previous first-order surveys.

A comparison of heights of the eleven bench marks as determined by instruments 85399 and 85394 shows the following differences:

Bench Mark	Quality Code	Diff.,mm	Dist., km
F 113	A	0.00	0.00
R 113	C	-2.01	0.47
ROD 1	B	-1.45	0.78

NBS 102	C	-1.20	1.14
L 113	C	-1.75	1.30
RV 6	C	-1.55	1.31
NBS 101	C	-1.95	0.95
RV 2	C	-3.00*	1.25
Q 113	A	-1.90*	1.25
RV 3	C	-0.89	1.49
NRC 3	A	-0.25	1.56

*Bench mark RV 2 was hit several times by lawn mowers and Q 113 was loose in the hole drilled in bedrock.

A running with instrument 85399 from bench mark RV 3 to Q 113 was 2.1 mm different from the other running, as well as from the mean of the two runnings of the other instrument. The section length was 247 m, measured in three setups. No reason was found for the larger than usual disagreement. Both runnings were used in the computation of heights.

One unusual setup, where sight lengths were about 27 m, should be described. When leveling on a gentle slope from NRC 3 to NBS 101, the instrument was set too low to read the forward rod at the horizontal crosshair. To experiment with the instrument, the "automatic read" was activated and a reading of -0.0192 m was determined by the instrument. Using this reading, the forward and backward running of this 300-m section agreed to 0.29 mm. There were no negative rod readings on the other running.

Based on the forward-backward differences (F+B), 12 of the 13 sections leveled agreed within first-order, class II tolerances, and one section each for both instruments agreed within second-order, class II tolerances. Based on loop misclosure tolerances, both surveys could be classified as first-order, class I.

HIGHWAY TEST

In order to evaluate the NA2000 system under actual field conditions, we releveled two sections of a first-order leveling line along State Highway 301 in Bowie, Maryland. These sections were double-run with Ni002 and MOM Ni-A31 levels in 1975-76. Highway 301 is a four-lane divided highway with heavy truck traffic. The temperature during the day was 28-35 degrees C.

The section from bench mark E 136 to L 135 is 1.36 km, the other section from L 135 to M 135 is 1.66 km. The forward run of the first section, 13 setups, took only 40 minutes and both instruments operated without any problems. But on the first setup of the backward run, one operator did not push the read button for the backsight; therefore the forward reading was recorded as a backsight reading. This transposition of readings was not detected until the 9th setup. For this running the backsight reading of the first setup was lost, and was estimated from the height difference measured by the other instrument for that setup and the foresight reading made by this instrument.

After the first section was completed, the observing party went to a nearby construction site where a bench mark had to be reset. Both instruments measured the height difference between the old and new marks twice from one setup with 36 m sight lengths. The height differences were: 0.5843 m and 0.5839 m for instrument 85394, and 0.5837 m and 0.5844 m for instrument 85399. In spite of the construction activities, heavy traffic, and sighting over asphalt, these differences agree within first-order, class II tolerances.

The second section was then leveled. The forward run, 16 setups, took 55 minutes. An attempt was made to complete the running in 15 setups, but due to interference from high grass around the bench mark, a reading was not possible, although the entire rod was visible in the field of view. During the backward run, on the 10th setup, one operator transposed the rod readings, i.e., the read/record button was accidentally activated and zero was recorded for backsight reading. The actual backsight reading was recorded as foresight and the foresight reading was recorded as backsight of the next setup. The following setups were also recorded similarly. All readings were recovered and the height difference for the section was recomputed. The observed height differences are:

Instrument	Bench Mark	Forward (m)	Backward (m)	F+B (mm)	Mean (m)
NA2000 85394	E 136-L 135	-8.38360	8.38469	1.09	-8.38414
	L 135-M 135	-0.28500	0.29489	9.89	-0.28994
NA2000 85399	E 136-L 135	-8.38240	8.38390	1.50	-8.38315
	L 135-M 135	-0.28560	0.28920	3.60	-0.28740
	Observations in 1975-76 were:				
Ni002	E 136-L 135	-8.38085	8.38055	-0.30	-8.38070
	L 135-M 135	-0.28897	0.28750	-1.47	-0.28824
MOM Ni-A31	E 136-L 135	-8.38367	8.38202	-1.65	-8.38284
	L 135-M 135	-0.28817	0.28690	-1.27	-0.28754

Three of four of the new measurements, based on B+F differences, agree within first-order, class I, and one agrees within second-order, class II tolerances. Comparing the NA2000 measurements to the mean of the 1975-76 measurements, all four agree within first-order, class I tolerances.

CONCLUSION

The NA2000 leveling system was designed basically for non-geodetic leveling surveys. The material and construction of the leveling rods do not satisfy FGCC specifications for precise leveling surveys. The computer software is structured to satisfy requirements for lower-order surveys; i.e., it computes height of instrument, heights of bench marks, heights of turning points, etc. If the rods were of similar construction to other precise leveling rods and the software were appropriately modified, this system could be used advantageously for precise leveling surveys. To be used most advantageously for motorized leveling, the instrument should have a swiveling eyepiece.

The system has many advantages:

1. A person is not required for recording because the instrument automatically records the rod readings.

2. Pointing and reading errors by the observer and recording and computing errors by the recording person are eliminated.

3. With this system the most difficult part of the observer's job, i.e., reading the rod, is automated; therefore the observer's attention can focus on other aspects of the observing process.

4. Longer sight lengths may be used on hot, sunny days because the "simmering" of the air does not seem to adversely affect rod readings, which translates into faster overall progress. Longer sight lengths are used only if the systematic refraction affect can be minimized by an appropriate correction.

ACKNOWLEDGMENTS

The author wishes to thank Mr. Rodney J. Lee for his assistance in data reduction and Mr. Gary M. Young for his valuable suggestions.

REFERENCES

Federal Geodetic Control Committee, 1984, Standards and Specifications for Geodetic Control Networks, National Geodetic Information Center, NOAA, Rockville, MD 20852, 33 pp.

Ellingwood, C.F. and Holdahl, S.R. 1972,"The Precise Leveling Test Network of the National Geodetic Survey," Preprint, Presented at the ACSM 32nd Annual Meeting, Washington, D.C., 13 pp.

Mention of a commercial company or product does not constitute an endorsement by the National Oceanic and Atmospheric Administration. Use for publicity or advertising purposes of information from this publication concerning proprietary products or the tests of such products is not authorized.

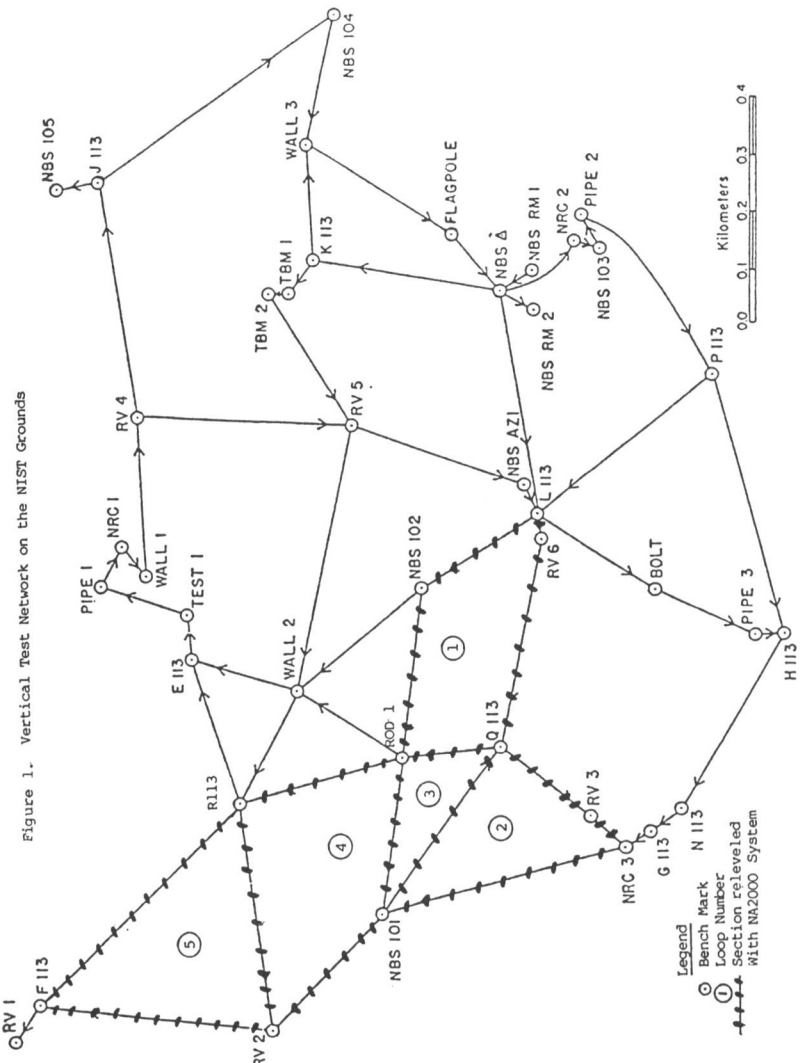

Figure 1. Vertical Test Network on the NIST Grounds

DESIGN CRITERIA FOR THREE-DIMENSIONAL COORDINATE COMPUTATIONAL SOFTWARE FOR FUTURE LAND INFORMATION SYSTEMS*

William E. Ball, Jr.

ABSTRACT

The most dependable three-dimensional coordinates obtainable of US Public Land Survey System (PLSS) corners and the most accurate representation obtainable of the actual ground surface will be needed for future land information systems, geographic information systems, topographic map construction, map revision, orthophotoplats with contour lines, and other applications that require three-dimensional coordinates of points on the earth's surface.

For many years conventional cadastral survey and computational methods and procedures have included simplifications and approximations based upon two-dimensional plane survey methods which will not be adequate to satisfy the three-dimensional land and resource management needs of the future.

To satisfy these future needs, three-dimensional cadastral survey computational software is being developed which is based upon the philosophy that there should be full compatibility among (1) the land surface, (2) coordinate systems, (3) survey measurements, (4) mathematical models and (5) computational theory. In addition, survey measurement data should be utilized as measured, without modification.

The criteria that are being applied in the development of the cadastral survey computational software, which are actually principles of good cadastral survey computational practice are described briefly.

INTRODUCTION

For future land and resource information systems (LIS), information will be needed from which an accurate representation of the actual ground surface can be obtained and from which accurate locations of land-ownership boundaries and resource-area boundaries can also be obtained.

Because of irregularities in the earth's surface (in the form of mountains, canyons, etc.), the land surface can only be represented accurately as a three-dimensional surface. Therefore, the locations of points on the surface should be defined for LIS applications in terms of three-dimensional coordinates. In addition, the locations of land-ownership and resource-area boundaries should also be defined in terms of coordinates.

Some resource-area boundaries, such as the boundaries of timber stands, for example, can be photographed and mapped reasonably accurately; and where vegetative cover is not excessive, ground surface elevations can be determined and contour lines constructed

---

* The opinions expressed are the author's and do not necessarily represent official policy.

reasonably accurately using photogrammetric mapping techniques – if in both cases sufficient accurate ground control is available.

When topographic maps can be obtained on which resource-area boundaries are accurately shown, and on which contour lines representing elevations are accurately located, three-dimensional coordinates of points can usually be determined by digitizing with sufficient accuracy for geographic information systems (GIS), land information systems (LIS), and similar applications.

In order to determine where land-ownership boundaries are located, and where resources are located in relation to land ownership, the location of land-ownership boundaries (such as US Public Land Survey System (PLSS) corners and gridlines) must also be determined with acceptable accuracy and in the same coordinate system as the resource-area boundaries.

Unfortunately, land-ownership boundaries such as PLSS grid lines cannot be photographed and mapped as easily and as accurately as more visible resource-area boundaries. Therefore, PLSS grid lines are sometimes incorrectly positioned on some maps from which coordinates are obtained by digitizing. Map location errors of PLSS corners of one-fourth mile or more have been discovered. In addition, some features of the PLSS grid such as 1/16 corners, some 1/4 corners, and the corners of many special surveys are not shown on maps. In such cases, the determination of land and resource ownership from maps can be no more accurate than the mapped locations of the PLSS grid lines.

Also, in some locations such as in heavily forested areas, some surface features might not be visible on photographs, and elevations might be difficult to determine photogrammetrically; in which case the accuracy and completeness of ground surface information shown on maps might not be adequate. Under these conditions, the accuracy and completeness of ground surface information obtained from maps can be no greater than the accuracy of the maps.

<u>Additional Land Management Information Could be Obtained from Cadastral Survey Data.</u> There can be no more factual information for map construction purposes and for land and resource management purposes than information obtained from accurate ground surveys and from field topographic observations. Potentially much valuable information could be obtained directly from and as a by-product of three-dimensional cadastral surveys.

Cadastral surveyors traverse hundreds of miles of public lands every year. The survey measurement data that is obtained, and the ground surface information that is – or could be – obtained by these surveyors could provide much of the information needed for topographic map construction, map revisions, land information systems, geographic information systems, orthophotoplats with contour lines, etc., on the public lands. However, official cadastral survey record data and data shown on existing cadastral survey plats is only two-dimensional. There is no information displayed from which terrain elevation and other ground surface information can be determined.

In current cadastral survey practice when modern surveying instruments are utilized, three-dimensional survey measurements (slope distances, vertical angles and horizontal angles) are obtained (or could be obtained easily). When obtained, this information is usually two-dimensionalized and otherwise modified, and two-dimensional rather than three-dimensional computations are performed (i.e., valuable information that is obtained - or could be obtained - is only partially utilized).

Conventional cadastral survey and computational methods include simplifications and approximations to compensate for past limitations in measurement and computational capabilities, and to save time and money. Typical examples of simplifications and approximations include the use of two-dimensional plane survey computational methods and mapping coordinate systems, which are only marginally adequate for large-area cadastral survey purposes.

In order to obtain more useable information for future land and resource management applications, developmental efforts in recent years have been directed toward the reduction or elimination of undesirable simplifications and approximations. Whenever possible an attempt has been made to obtain more information from survey data without unnecessary additional field survey or computational effort.

The cadastral survey computational software that has been developed in recent years has been designed to utilize appropriate modern surveying instruments, suitable existing computational equipment, and theoretically correct analytical concepts. Corresponding compatible field surveying and computational methods and procedures are also necessary to assure maximum compatibility among instruments, equipment, theory and practice.

Some of the cadastral survey data that is produced can be used directly for LIS applications, and some will be used indirectly as less-than-control-survey-quality ground control for photogrammetric mapping and for other map construction and map revision applications from which useable LIS information can then be obtained by digitizing.

The criteria that have been applied in recent surveying computational software development, which is the subject of of this presentation, are actually principles of good cadastral survey computational practice. Some of the most significant of these principles are listed below, and are then described briefly.

SOME PRINCIPLES OF GOOD PRACTICE FOR THREE-DIMENSIONAL CADASTRAL SURVEY COMPUTATIONS

1. Survey measurements (as obtained), mathematical models, computational theory, and coordinate systems utilized should all be compatible and should conform to the geometry of the actual land surface.

2. The natural coordinate system of a surface (or an exact mathematical transformation of the natural coordinate system) should be used for geometric computations associated with the surface.

3. A three-dimensional geometric approach should be utilized for computations associated with three-dimensional surfaces.

4. <u>Exact</u> computational methods should be utilized for large-area cadastral survey computational purposes.

5. Survey computational and error adjustment capabilities should be sufficiently comprehensive and flexible to cover the wide variety of actual conditions and types of errors that might reasonably be encountered.

6. Adequate computer hardware and properly designed software should be utilized to assure the retention of computational accuracy throughout the computational processes.

These principles are recommended for all large-area cadastral survey computations and for the development of all new cadastral survey computational software.

A large-area cadastral survey is defined as the survey of a parcel of land of such size that the accuracy of the survey would be affected adversely without proper consideration of the effects of earth curvature and terrain elevation.

It is assumed that the three-dimensional software will be used for the primary purpose of obtaining accurate computational results. Therefore, in all recently developed software, there is a basic set of objectives – that the scientific computational capabilities developed should be (1) theoretically correct, and (2) as simple to use as practicable. When both objectives cannot be satisfied, accuracy and correctness are given priority.

A brief description of each of the principles is included in the following sections.

BRIEF DESCRIPTIONS OF THE PRINCIPLES OF GOOD COMPUTATIONAL PRACTICE

<u>Survey measurements (as obtained), mathematical models, computational theory, and coordinate systems utilized should all be compatible, and should conform to the geometry of the actual land surface.</u>

Because the earth's ellipsoidal shape must be properly considered for large-area cadastral survey purposes, and because surface topography must be correctly considered for land and resource management purposes; the actual land surface must be represented as realistically as possible – as a highly irregular three-dimensional surface.

When a cadastral survey is performed utilizing modern surveying instruments and techniques, three-dimensional measurements including slope distances, vertical angles, and horizontal angles among points on the actual land surface are, or can be, easily obtained. If, in addition, the survey measurements are properly oriented (i.e., oriented with respect to true north and vertical) then the measurements not only correspond to the actual surface, they also correspond to the geographic coordinate system – i.e., true north, true east, and vertical measurements correspond to increments in latitude, longitude, and elevation. Also, when modern positioning equipment is utilized, three-dimensional geographic coordinates of points on the actual land surface are obtained. These measurements and coordinates represent the actual surface, and they should be used as measured without modification. In addition, throughout all subsequent computational processes the true land surface locations, relative positions, shapes, sizes and orientations as defined by the survey measurements and observed coordinates should be preserved. This can be accomplished by (1) utilizing a proper coordinate system that conforms as nearly as possible to the actual land surface; (2) utilizing a mathematical model that accurately represents the actual land surface; and (3) applying computational methods that properly and accurately simulate the three-dimensional shape of the land surface, utilize the survey measurements, observed coordinates and orientations as obtained without modification, and utilize the properly chosen coordinate system.

Under these conditions full compatibility exists. All properly oriented field measurements can be utilized as measured without reductions or corrections for directional orientation, elevation, scale, etc; and the effects of earth curvature, convergence of meridians, etc. are properly provided for automatically as a natural benefit of the three-dimensional ellipsoidal geometry. In addition, computational equations can be expressed in their most direct and most exact forms. Also, because corrections to compensate for incompatibility among coordinate systems, survey measurements, and the actual land surface are not needed; the number of computational operations is reduced, chances of error are minimized, etc.

In recent years, nearly all cadastral survey computational software has been designed to utilize either full three-dimensional survey measurement data as measured in the field, or data that has been partially reduced to an approximate three-dimensional surface at average terrain elevation as displayed on a cadastral survey plat. In addition, all computer programs have utilized three-dimensional geometric concepts based upon the appropriate mathematical model of the earth's surface, and have been developed utilizing geographic coordinates. In this way 3-D survey computational systems developed have been designed to achieve maximum practicable compatibility.

<u>The natural coordinate system of a surface (or an exact mathematical transformation of the natural coordinate system) should be used for geometric computations associated with the surface.</u>

The natural coordinate system of a surface is a coordinate system that conforms to the actual shape, size, and orientation of the surface; or in the case of a highly irregular surface, the natural coordinate system is one that conforms to a mathematical model that represents the true surface as closely as possible for computational purposes.

In North America over a period of many years, the earth's surface has been represented by several reference ellipsoids (or spheroids). The natural coordinate system of these ellipsoids is the geographic coordinate system; the coordinates of which are expressed in terms of latitude, longitude and elevation (ϕ, λ, h). Since the geographic coordinate system is the natural coordinate system of the ellipsoids that most accurately represent the land surface, the geographic coordinate system or a coordinate system that is an exact mathematical transformation of the geographic coordinate system should be used for geometric computations associated with the actual land surface.

Utilization of Transformations of Natural Coordinates - In some cases computations can be greatly simplified with no approximations and no loss of accuracy using coordinate systems for computational purposes that are an exact mathematical transformation (not a map projection) of the natural coordinate system. For example, the straight line distance between two points in space (as might be measured using an electronic distance measuring instrument unaffected by refraction) can be computed in less time and with less computational effort in terms of a proper set of three-dimensional cartesian (x,y,z) coordinates than in terms of latitude, longitude, and elevation (ϕ,λ,h). In such cases, an exact transformation from ϕ,λ,h to tangent plane (or tangent space) x,y,z coordinates or to geocentric x,y,z coordinates can be performed, and the straight line distance can be computed in terms of x,y,z. Because straight line distance is an invariant under coordinate transformation the distance computed in terms of x,y,z is identically the same as the distance computed in terms of ϕ,λ,h.

Mapping Coordinate Systems Should Not be Used for Cadastral Survey Computational Purposes - Because mapping coordinate systems are based upon deformed (flattened) surfaces, all true directions must be adjusted to obtain grid directions and all true distances must be adjusted for the effects of elevation differences and for differences in scale. In addition, adjustments for the effects of earth curvature and the convergence of meridians are required. All of these adjustments vary with location and, therefore, must be determined at each point. For these reasons, mapping coordinate systems should not be used for large-area cadastral survey or computational purposes. (One only has to observe the distorted shape of Greenland on a map compared with the shape of Greenland on a globe to see how a deformed mapping coordinate system compares with an undeformed ellipsoidal coordinate system.)

Natural coordinate systems have been utilized in all cadastral survey computational software that has been developed in recent years. Either geographic coordinates or exact transformations - such as tangent space (x,y,z) coordinates (when appropriate) - have been utilized in all software developed for three-dimensional survey applications.

For two-dimensional cartographic and graphics applications when flattened surfaces are desired, applicable UTM, state plane, and other types of mapping coordinates have been utilized. In some cases, two-dimensional tangent plane (x,y) coordinates have also been utilized for two-dimensional applications.

A three-dimensional geometric approach should be utilized for computations associated with three-dimensional surfaces.

When an accurate and realistic representation of the actual land surface is needed for land and resource management purposes and/or for large-area cadastral survey purposes, the earth's surface must be treated for computational purposes as though it was a highly irregular surface in three-dimensional space. When properly oriented three-dimensional survey measurements in the form of slope distances, horizontal angles, and vertical angles have been obtained; and when geographic coordinates of points on the actual land surface have been obtained; then sufficient information is available to accurately define the shape, size, and location of a parcel of land which lies on the actual surface.

Under these conditions and in order to preserve the true land surface locations, relative positions, shapes, sizes, and orientations throughout the computational processes a three-dimensional computational approach must be applied that accurately simulates the actual shape of the earth and which will properly consider the effects of earth curvature, convergence of meridians, elevation above sea level, three-dimensional surface topography, etc. The three-dimensional computational system should also utilize the three-dimensional survey measurement data and observed point position coordinates as measured without modification. Upon completion the computational results should be output in the form of geographic coordinates, distances at terrain elevation, true bearings, etc.; i.e., as the actual unmodified results of a proper set of three-dimensional geometric computations.

In the past, two-dimensional computational methods based upon plane geometric techniques and mapping coordinate systems (such as state plane coordinates) were utilized for land survey computations to save time, reduce computational costs and/or to compensate for limitations in surveying and computational capabilities. As a result of modern technological advancements, especially in surveying instruments and in the power, speed and availability of computers and modern software there is no longer any need to utilize two-dimensional methods. Two-dimensional methods can now be replaced with modern three-dimensional computational software that is theoretically correct, utilizes survey data as measured without modification, and which will produce accurate results at little or no greater cost than the two-dimensional methods.

Software that utilizes a three-dimensional geometric approach has been developed in various forms which can be used with three-dimensional coordinates as observed and with three-dimensional survey measurement data as recorded, or which can be used with data that has been partially reduced to an approximate three-dimensional surface at average terrain elevation as might appear on a cadastral survey plat.

<u>Exact computational methods should be utilized for large area cadastral survey computational purposes.</u>

Exact computational methods utilize mathematical equations that (1) accurately represent a set of actual conditions, or (2) represent an accurate mathematical model of a set of actual conditions.

Approximate computational methods utilize simplified mathematical equations that (1) approximately represent a set of actual conditions, or (2) represent a simplified mathematical model of a set of actual conditions.

Exact computational methods produce accurate results over a broad range of conditions and utilize survey measurements as observed without modification.

Approximate computational methods can only be depended upon to produce useable results over a limited range of conditions, or can be depended upon to produce useable results only when appropriate correction factors have been applied. Computational methods that utilize plane coordinate computational techniques, for example, which might be useable for computations associated with small lots in limited areas might not be useable for cadastral survey computations associated with large areas that are affected by earth curvature, convergence of meridians, elevation differences in irregular terrain, elevation above sea level, etc. Also, in the case of approximate methods, actual survey measurements must sometimes be modified for compatibility with a simplified or distorted mathematical model or inappropriate coordinate system.

Exact mathematical methods developed for large area computations in three-dimensional space can usually be utilized safely for small area computations in two-dimensional space, but approximate mathematical methods adequate for small area computations in two-dimensional space cannot always be used safely for large area computations in three-dimensional space.

With extremely fast, powerful and accurate existing computing equipment accessible to all surveyors, and with software based upon exact mathematical methods readily available there is no longer any need to use approximate methods for large area cadastral survey computational purposes.

Nearly all recently developed cadastral survey computational software has utilized a geometrically exact mathematical approach whenever practicable.

<u>Survey computational and error adjustment capabilities should be sufficiently comprehensive and flexible to cover the wide variety of actual conditions and types of errors that might reasonably be encountered.</u>

The complexity and diversity of conditions that a cadastral surveyor might encounter is nearly limitless. There is no single computational approach that can be used to obtain meaningful results under all possible circumstances. Conditions exist that require many different computational approaches, chronological sequencing, hierarchial sequencing, etc.

Maximum reasonable automation should be a major objective in the development of new software, but correct results must be the primary objective. When both objectives cannot be achieved equally, then correct results must be given priority. Scientific software should be designed to save time and labor, but not to make important technical decisions automatically that require professional analysis and judgement. If full automation can only be achieved by sacrificing accuracy or correctness then full automation cannot be justified.

Full automation can only be achieved when needed computational sequences can be predicted in advance. Fully automated systems can only be justified economically when computational sequences are repetitive. Because the US Public Land Survey System is so diverse and so complex, computational sequences are rarely predictable or repetitive.

Fully automated systems can and should be developed to handle a finite number of more or less standardized and/or predictable options, but when an infinite number of possible computational and error adjustment conditions exist, some of which cannot be automated safely, it sometimes becomes necessary for a knowledgeable surveyor to control the computational and/or adjustment processes.

Complex problems are frequently a combination of simpler problems that can be solved most easily and most correctly by a knowledgeable person who understands both the computational theory and the survey problem well enough to separate a complex problem into its simpler components. The solution to the complex problem then becomes a combination of solutions to a set of simpler problems. If a fairly complete set of programs is available then a very large variety of complex problems can be handled.

Different types of errors are also likely to occur, and there is no one type of adjustment that can be used to adjust all types of errors correctly. A preprocessing capability can be used to assist in detecting, identifying and locating different types of errors, but the proper adjustment procedures might not be predictable in advance as would be required for a fully automated system. Blunders can sometimes be isolated, but whether they can be corrected legally cannot be predicted in advance (as would legally permissible in the case of a data entry error, for example). Systematic errors might be modeled and then adjusted using an appropriate rule of adjustment. Random errors might be adjusted using a least squares approach, etc; but the best adjustment approach might sometimes be difficult to predict in advance for purposes of automation.

For these reasons a comprehensive survey computational system or software package must include an extensive assortment of computational and adjustment options. The system should also provide automated capabilities for relatively simple, predictable situations and manually controlled capabilities for more complex situations.

Current cadastral survey computational software includes an extensive system of computational and adjustment options which when combined properly will solve a very large number of computational and adjustment problems. Data and error evaluation programs have been included to assist in determining the proper computational and/or adjustment approach to be applied. In addition, both automated and manually controlled capabilities have been provided that can be utilized in an appropriate manner by a knowledgeable surveyor. This provides maximum computational and adjustment flexibility, but requires knowledgeable, experienced and conscientious personnel.

<u>Adequate computer hardware and properly designed software should be utilized to assure the retention of computational accuracy throughout the computational processes.</u>

Much has been said in scientific literature concerning the total number of significant figures and the number of significant figures following a decimal point that are needed to define the constants, measurements, and variables to a required degree of accuracy that are utilized for scientific applications.

Much less has been said concerning the precautions that should be taken to retain the needed level of accuracy throughout a series of computations. Beginning with accurately measured and accurately defined information is essential, but doesn't guarantee that computational results of equivalent accuracy will be obtained.

Loss of accuracy due to such causes as the use of computers with insufficient precision, the improper utilization of two-dimensional solutions or approximate solutions instead of exact solutions, the utilization of inappropriate coordinate systems, etc., can be avoided if there is an awareness that these and other causes can be significant sources of error under some circumstances. Several of these potential causes of loss of accuracy have been discussed in previous sections.

Loss of accuracy due to insufficient precision can be avoided by observing the "2n Rule of Significant Figures" when selecting a computer for scientific computational purposes. According to this rule a computer must be capable of storing temporarily a numerical value having two times as many significant figures as the number of significant figures of numerical accuracy that must be retained.

Ten or eleven digit numbers occur frequently in surveying, mapping, geodesy and photogrammetry. For example, the earth radius is 20,925,832.164 ft., 200 nautical miles is 1,215,220.667 ft., a longitude of 100°00'0.001" expressed in degrees to the nearest 0.1 ft. in position would be 100.00000028 . In these cases n = 10 or n = 11 and 2n = 20 or more. Therefore, according to the 2n Rule of Significant Figures, a computer capable of storing temporarily a 20 significant figure number should be used to avoid any computational loss of accuracy (during multiplication.)

In the case of a 32 bit computer there are approximately fourteen significant figures in a double precision word and twenty eight or more significant figures in a quad precision word. If, according to the 2n Rule of Significant Figures, 20 significant figures of precision are required to prevent computational loss of accuracy; double precision would be insufficient. In this case quad precision should be selected.

(Quad precision is not normally available in PCs, but there are many small computers available that were designed for scientific purposes which are physically not much larger in size than a PC and which provide quad precision and other desirable scientific computational capabilities).

When all computations have been completed the results are then rounded according to the familiar rules for displaying numerical results that are consistent with the accuracy of the constants and measurements. Therefore, quad precision is only needed for computational purposes, not for permanent file or database purposes.

Recently developed cadastral survey computational programs have been designed to utilize computing equipment that will provide the word size, speed and accuracy needed to prevent computational loss of accuracy, and which will also provide other desirable scientific computational capabilities.

STATUS OF THREE-DIMENSIONAL COMPUTATIONAL SOFTWARE

Development of the three-dimensional survey computational software has followed a four-stage evolutionary process on a time-available basis which has spanned a period of several years. In the first stage of development mathematical equations were derived to provide solutions to a wide variety of three-dimensional computational tasks. Those equations were then coded to produce a number of small independent programs which could be used to verify the correctness of each of the tasks. The small programs were subsequently utilized as subroutines within larger more comprehensive programs which, when combined with crude data input/output capabilities, produced stand-alone programs that could be tested for accuracy and completeness. In the second stage those programs were integrated using simple file management capabilities which would transfer data among the programs thereby creating a somewhat crude but functional prototype system upon which systems testing could be performed. The third stage will include more efficient data entry capabilities, more efficient file management capabilities, and potential-user testing of the system as a whole for computational efficiency and functional effectiveness. The fourth and final stage should produce a correct, efficient, documented, and productive operational system.

The three-dimensional computational software is nearing the end of the second, i.e., prototype, stage of development.

SUMMARY

The most accurate representation obtainable of the actual land surface, land-ownership boundaries, and resource-area boundaries will be needed for future land information systems, geographic information systems, topographic map construction, map revision and other applications that require dependable information associated with points and lines on the earth's surface.

The most factual information obtainable for these purposes is the survey data and topographic observations that can be provided by cadastral surveyors who traverse hundreds of miles of public lands every year.

In anticipation of the need to extract as much land surface information as possible from cadastral survey data three-dimensional cadastral survey computational software is being developed. This software has been designed to achieve maximum compatibility among (1) the land surface, (2) coordinate systems, (3) survey measurements – as observed, (4) mathematical models, and (5) computational theory.

The design criteria that have been applied in the development of the software and the current status of the software have been described.

A COMPARATIVE EVALUATION FRAMEWORK FOR CADASTRAL SYSTEMS IN THE U.S.

Grenville Barnes
Department of Geodetic Science and Surveying
The Ohio State University
Columbus, Ohio 43210-1247

Abstract

Support for the Multipurpose Cadastre concept as identified by the National Research Council Reports (NRC 1980; NRC 1983) in the early eighties appeared to have waned by the second half of the decade. However, as more counties automate their records and develop computerized cadastral systems, questions as to approach and design have once again come to the fore. This paper focuses on the cadastral overlay component of a Multipurpose cadastre or LIS and addresses the question facing many local governments: what is the best way to create a cadastral overlay?

In developing an evaluating framework, one must begin by understanding what is meant by the term "best" system and, more importantly, the fact that this will vary between different users, environments, and cultures should be taken into account. From a surveying standpoint, we have tended to regard the "most accurate" as "the best" even though this accurate approach may the "worst" in terms of cost, flexibility, and other factors. This paper describes an evaluation framework which broadens traditional evaluation approaches by considering a set of several criteria which includes accuracy, cost, flexibility, utility, maintainability, and simplicity.

The underlying message behind this paper is that there is very rarely an ideal approach which maximizes all criteria, and there will almost always be a certain "tradeoff" between these criteria.

SPHERICAL TRIGONOMETRY - KEY TO UNDERSTANDING WORLD GEOGRAPHY, NAVIGATION AND CONTROL SURVEYS

Dexter M. Brinker, PE-PLS
Consulting Engineer and Land Surveyor
P.O.Box 3092, Durango, Colorado 81302

ABSTRACT

For some reason, spherical trigonometry has been dropped from the curricula of most high schools and colleges. I suppose this results from the fact that a few knowledgeable people can provide enough spherical trigonometry computer software to supply the calculation needs of the nation, and all the average user needs to know is how to enter sufficient data into the "black box" along with a request for desired answers.

I have no desire to belittle the ability of computers to solve routine mathematical calculations faster than the human brain, but I feel that the users of these computers should have some idea of what is going on inside the box, lest the user become a slave of the machine.

In the case of spherical trigonometry, as applied to navigation, astronomy, control surveys, map projections and geographic information systems, an understanding of the basic concepts and the ability to derive at least a few basic equations can help elevate a person from the status of a technician to that of a professional and also provide a great amount of self satisfaction.

DISCUSSION

When we rotate a circle $180°$ around any of its diameters, the circle generates a surface we call a sphere. This diameter of rotation is called the axis, and for geographic and astronomic work, one end of the axis is called the north pole and the other end the south pole. It is traditional, at least in the northern hemisphere, to put the north pole at or near the top of drawings portraying the sphere. One of the most useful ways of representing a sphere on a flat piece of paper is by means of the orthographic projection. (See "Descriptive Geometry of the Celestial Sphere", A.C.S.M. Bulletin, April 1986 and October 1987.)

When a plane cuts through a sphere, the intersection is a circle. If the plane passes through the center of the sphere, the circle is called a "great circle" and it divides the sphere into two hemispheres. If the cutting plane does not go through the center of the sphere, the intersection is a "small circle". In surveying and geography, the most important small circles are circles of latitude, formed by cutting planes which are above or below the equator and perpendicular to the axis. The equator at $0°$ latitude is, of course, a great circle. In astronomy, the counterparts of latitude circles are called declination circles.

A part of a spherical surface which is bounded by three great circle arcs is called a spherical triangle, and it is the solution of this type of triangle that forms the basis of spherical trigonometry. Spherical triangles are different from plane triangles in that both the sides and angles are measured in terms of angle measurement, usually degrees. Another important difference is that while the interior angles of a plane triangle always total $180°$, the sum of the interior angles of a spherical triangle is more than $180°$ but less than $540°$, depending on how much of the surface of the sphere it covers.

For geographic work and navigation, one of the most useful spherical triangles is the one formed by longitude arcs from the pole to two points on the sphere and the arc between the two points often called the "great circle route". For example, if we know the latitude and longitude of two points on the globe, as in Figure 1, we can, with appropriate equations from spherical trigonometry, compute the angular distance between the two points and the bearing of each point from the other. The angular distance can be readily converted to nautical miles or statute miles.

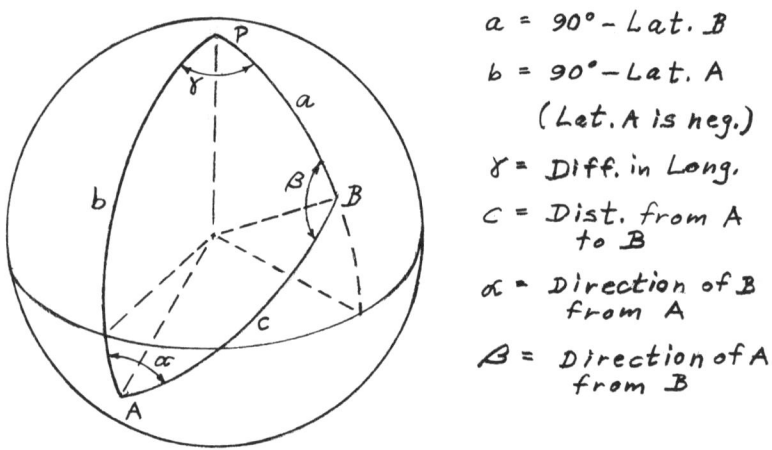

$a = 90° - Lat. B$

$b = 90° - Lat. A$

(Lat. A is neg.)

$\gamma =$ Diff. in Long.

$c =$ Dist. from A to B

$\alpha =$ Direction of B from A

$\beta =$ Direction of A from B

$$\cos c = \cos a \cos b + \sin a \sin b \cos \gamma$$

$$\sin \alpha = \frac{\sin a \sin \gamma}{\sin c} \qquad \sin \beta = \frac{\sin b \sin \gamma}{\sin c}$$

Figure 1

For surveys of relatively small areas, the spherical nature of the survey surface is usually ignored and the techniques of plane surveying and plane trigonometry are employed. However, as the extent of a survey increases, as in a control survey for a large county, the curvature of the earth must be reckoned with in order to maintain appropriate precision and overall accuracy. In order to know when the earth's curvature is a problem, we make use of an interesting characteristic of "spherical excess", that is, the amount by which the sum of the interior angles in a spherical triangle exceeds 180°. The ratio of this quantity to 360° (the limiting amount of spherical excess) is the same as the ratio of the area of the spherical triangle to the area of a hemisphere having the same radius as the spherical triangle. In Figure 2 we have two longitude arcs on the earth extending from the pole to the equator. The arcs are one second apart and the third side of the spherical triangle is a short arc on the equator which is perpendicular to both longitude arcs. Assuming the radius of the earth to be about 3960 miles and using the relationship that the area of a sphere is $4\pi R^2$, we see that the area of this sliver is:

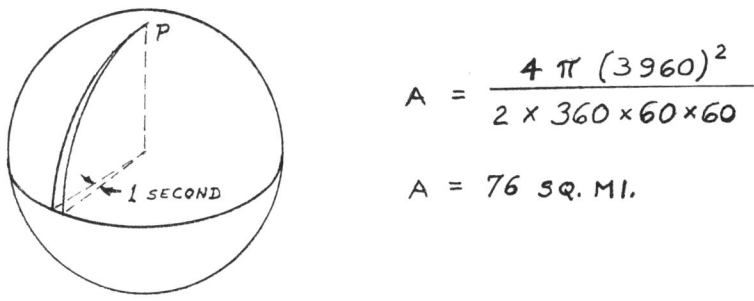

$$A = \frac{4\pi(3960)^2}{2 \times 360 \times 60 \times 60}$$

$$A = 76 \text{ SQ. MI.}$$

Figure 2

Since the sum of the interior angles in the triangle is 180° 00' 01", the spherical excess is one second and we can conclude that if a triangle in a control survey network has an area of 76 square miles, the observed interior angles should total 180° 00' 01", not 180° 00' 00" as in a plane triangle. Larger triangles have proportionately larger spherical excess.

When using state plane coordinates, the complications of spherical trigonometry are avoided by applying small angular corrections, known as "second term" corrections, at each observation station. These are necessary because when a long geodetic line, such as the side of a spherical triangle in a control survey, is shown on a state plane grid, it is concave toward the central parallel (Lambert) or central meridian (transverse Mercator). Figure 3 shows a control system triangle in the Lambert system with sides of approximately 24, 32 and 40 miles. The broken lines represent the "map images" of the geodetic lines, and each is concave toward ϕ_0, the central parallel for the zone. The angle formed at each station by tangents to the geodetic lines (e.g. ∠ B'BB") represents the <u>observed</u> angle.

For each triangle these angles should total 180° plus the spherical excess. The small angles between the sides of the plane triangle ABC and the tangents to the curved (geodetic) lines are the second term correction angles, and must be calculated for both ends of each line using equations furnished with state plane coordinate tables. The algebraic sum of the second term corrections equals the spherical excess which can be determined accurately enough from the area of the plane triangle. In the example this excess would be about 5 seconds. A sketch similar to Figure 3 should be used to assure that the corrections are made in the right direction. The forward and back grid bearings on any line are numerically equal, but the forward and back geodetic bearings usually are not, and depend on both the mapping angle (difference between grid north and geodetic north) and the second term corrections at each end of the line.

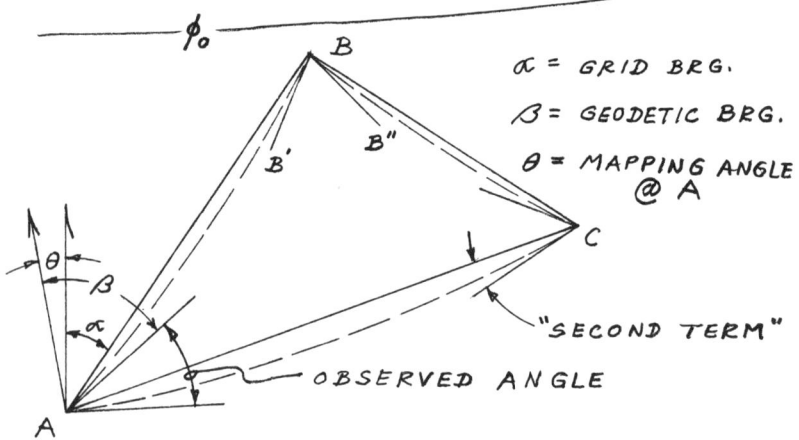

Figure 3

In astronomic work, a spherical triangle similar to that used in terrestrial applications is made up of great circle arcs on the celestial sphere. The first arc extends from the pole to the observer's zenith, the second from the pole to the celestial body being observed, and the third from the zenith to the celestial body. This is the well known PZS triangle. One more great circle is usually added, namely the observer's horizon, which is obtained by passing a plane through the center of the celestial sphere and perpendicular to the radial line from the center to the zenith. In a practical sense, this plane should pass through the observer's station and be tangent to the earth, but the offset (parallax) is negligible for all normal surveying applications except for observations on the sun, in which cases suitable corrections are made. The various parts of the celestial sphere and PZS triangle are shown in Figure 4. By measuring certain angles of the PZS triangle (or in some cases, the complements of these angles) we get sufficient data to calculate the quantity we want. Two variations of the measurement process should be noted. The first is that often we accept measurements made by someone else; e.g. the declination of the sun or stars. Secondly, angles related to longitude are often measured not by a transit or theodolite, but by a clock. This is possible because of the uniform speed of rotation of the earth.

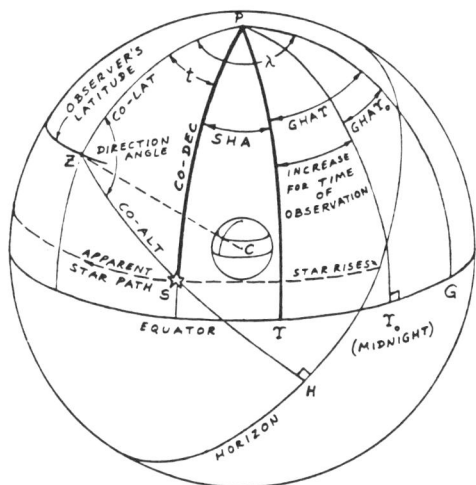

Figure 4

Latitude circles, as used in the public land survey system, present a special application of spherical trigonometry. If a surveyor sets up a surveying instrument somewhere between the equator and the north pole, determines north and turns an angle of 90°, he will be looking east or west. However, if he starts to prolong a "straight" line, it will actually be a great circle arc which would cross the equator if it were possible to extend the line a quarter of the way around the earth. In order to stay "on course", the surveyor would, after going a few miles, have to solve a very slim right spherical triangle and determine the proper distance to go north in order to get back to the starting latitude. (See Page 54, Manual of Surveying Instructions 1973, Bureau of Land Management.) This situation is shown in exaggerated form in Figure 5.

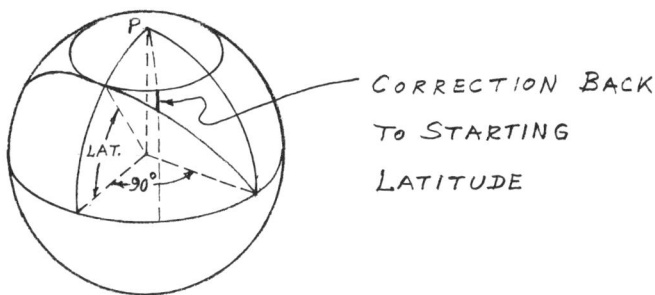

Figure 5

CONCLUSION

As the use of global positioning systems becomes more common and geographic information systems extend over ever greater areas, it is essential that surveyors and mappers understand the importance and power of spherical trigonometry. In addition, many other factors need to be taken into consideration when planning and executing large area surveys. These include adequate instrumentation and field techniques, repetition of measurements, elimination of blunders and proper management of errors. The techniques of plane surveying and plane trigonometry must be reserved for only those surveys of very limited extent.

BLUNDER DETECTION IN A SMALL GEODETIC NETWORK

John M. Brown
Chevron Exploration and Production Services Co.
Navigation and Survey Services
Houston, Texas 77242

James S. Bethel
Purdue University
School of Civil Engineering
West Lafayette, IN 47907

ABSTRACT

The motivation for this paper came from a desire to encourage more widespread use of powerful statistical methods for detecting and locating blunders in observed data. Such methods often remain theoretical exercises instead of practical tools for everyday use. Four methods were chosen for evaluation in a fictitious triangulation/trilateration network with simulated random errors and isolated gross errors introduced. The four methods analyzed are (1) data snooping, (2) tau statistic, (3) iteratively reweighted least squares, and (4) L1 norm minimization. A scoring system was devised to evaluate the effectiveness of each technique in correctly locating the observations with blunders.

INTRODUCTION

Blunder detection is an essential component of data analysis and adjustment. Without elimination of gross errors, the results of traditional least squares adjustment become very questionable. Following a conventional least squares adjustment, an F test or Chi-square test on the a posteriori reference variance at usual significance levels will not necessarily indicate the presence or absence of gross errors. It is therefore often recommended, and we concur, that some method of blunder detection should be employed regardless of the outcome of the global reference variance test. Since computational power is today so readily available, there are few reasons to defer the use of such methods. This paper focuses on four of these methods: (1) data snooping, (2) tau statistic, (3) iteratively reweighted least squares, and (4) L1 norm minimization.

BLUNDER DETECTION METHODS

Data Snooping

The method commonly referred to as data snooping was originally developed by Baarda and is based on the multivariate statistical theories of Scheffe. In essence it performs a statistical test on the standardized residuals of the observations. An important condition required here is that the variances of all observations must be known a priori. The method assumes that only one observation at a time is corrupted and is thus a univariate test. Given a diagonal weight matrix, in other words, no correlation between the observations, the following statistic may be written:

$$w_i = \frac{v_i}{(\sigma_0 * \sqrt{qvv_{ii}})} \qquad (1)$$

where v_i is the ith residual, σ_0 is the a priori reference variance, and qvv_{ii} is the ith diagonal element of the variance/covariance matrix of residuals. A critical value, c, is calculated from:

$$c = \sqrt{F((1-\alpha),1,\infty)} \qquad (2)$$

where F denotes the abcissa of the F-distribution at the given alpha level. Baarda suggests the value for α to be 0.001. This yields a value for c of 3.29. Thus any residual for which the test statistic w_i exceeds 3.29 will be flagged as a blunder. In practice multiple blunders could be handled by successive reapplications of this test, with observations eliminated as flagged after each step.

Tau Statistic

In the case that that the observation variances are not known a priori, then the above method may not be used and the corresponding test is based on the tau statistic. This method was developed by Pope (1976) and is based on the tau distribution which is derived from the t distribution. The test statistic is calculated very similarly to the previous case, with the a posteriori reference variance being used instead of the a priori value. Again, given a diagonal weight matrix the test statistic is:

$$\tau_i = \frac{v_i}{\hat{\sigma}_0 * \sqrt{qvv_{ii}}} \qquad (3)$$

where v_i and qvv_{ii} are as described above, and $\hat{\sigma}_0$ is the a posteriori reference variance. Pope supplies a subroutine to calculate critical values from the tau distribution. The blunder detection/location strategy proceeds in a similar fashion as the previous method. Off-diagonal elements of Qvv may be useful in order to examine and interpret residuals which are highly correlated. This however greatly increases the cost of the analysis.

Iteratively Reweighted Least Squares

This method has also been referred to as robust estimation. It was introduced in the mid-twentieth century and adapted to geodetic networks a few years later (Krarup, 1980). In this method, the weights of the observations are treated as dynamic quantities. They are allowed to change with each iteration of the solution. Observations with high residuals are given low weights, and those with small residuals are held stable. When the solution converges, the weights of the outlying observations have approached zero, thus effectively removing those observations from the adjustment. No particular changes are necessary to handle the case of multiple gross errors.

As each iteration progresses, observations must be allowed to move "in and out" of the adjustment. Selecting the weight function seems to be more of a trial and error operation than a strictly theoretical one. In our case we followed the recommendation of Krarup (1980) and used the following weight functions:

$$1st\ Iteration: w_i = w_i \qquad (4)$$

$$2nd,\ 3rd\ Iteration: \text{if } \left|\frac{v_i}{\sigma_i}\right| > 3.0,\ then\ w_i = \left[e^{-\left|\frac{v_i}{\sigma_i}\right|^{4.4}}\right]^{0.05} \qquad (5)$$

$$4th + Iteration: \text{if } \left|\frac{v_i}{\sigma_i}\right| > 3.0,\ then\ w_i = \left[e^{-\left|\frac{v_i}{\sigma_i}\right|^{3.0}}\right]^{0.05} \qquad (6)$$

where w_i is the weight of the ith observation, v_i is the ith residual, and σ_i is the a priori standard deviation of the ith observation.

There are several advantages to this method and some drawbacks. A computational advantage is that the matrix Qvv need not be calculated. A second advantage is that when the solution converges and the outliers have weights near zero, their corresponding residuals reflect the gross error magnitude more accurately than residuals from conventional least squares. The same information on the gross error magnitude is obtainable with the earlier discussed methods, but requires recomputation without the observations in question. No rigorous hypothesis testing can be done with this method, rather some multiple of the a priori standard deviation of each observation can be used as a critical value.

A disadvantage is that, at least during the course of these experiments, numerical problems were encountered in solving the system of equations when large blunders and high correlation among the residuals existed. One is forced to stratify the gross error magnitudes and assume that very large gross errors have been eliminated by some preliminary screening method.

Minimization of the L1 Norm

The last technique used in the experiment was adjustment by minimization of the L1 norm instead of the L2 norm as with conventional least squares. This can be expressed as:

$$\sum_{i=1}^{i=n} |v_i| \rightarrow \min \tag{7}$$

In other words, the sum of the absolute value of the residuals is used as a minimizing criteria rather than the sum of the squares. This can be likened to taking the median of a set of observations rather than the mean. Rather than using a post-adjustment technique as the previous three were, this technique can be used as a pre-adjustment to detect and locate possible outlying observations and eliminate them from the data set before actual adjustment is performed. This method has probably been avoided in the past because the solution becomes a linear programming problem and such techniques are not widely used by geodesists. However, simple and efficient algorithms have been developed which make the technique fairly accessible. This study used an algorithm based on a revised simplex tableau (Barrodale and Young, 1966).

As in the case of iteratively reweighted least squares, no formal test statistics are computed to determine if an observation is an outlier. The decision is made by comparing the residuals with some multiple of the a priori standard deviation. No numerical problems were encountered with this method.

TEST NETWORK

In order to evaluate the effectiveness of each of these techniques, a small surveying network (2D) was developed consisting of 26 observations, four fixed control points, and four unknown points. The network was based on an actual survey that was slightly modified. See Figure 1. for a sketch of the network. Perfect observations were calculated and then perturbed with random errors, $\sigma = 0.05$ feet for distances and $\sigma = 5$ seconds for angles. To evaluate the effectiveness of each technique, intentional blunders were introduced into the network one observation at a time and in all possible combinations of pairs. A small program was written that looped 26 times for single observations, and 650 times for pairs (each pair occurs twice). Each time an observation was perturbed by a predetermined amount. Blunders introduced into the network were multiples of the a priori standard deviation: 4x, 5x, 10x, 20x, 50x.

All techniques started out with a score of zero. If the method caught an outlier at the correct time, it received one point. If it missed the outlier it received zero points. If it incorrectly flagged an observation as an outlier one point was deducted. If for example only the sixth observation had been perturbed but the sixth, eighth, and ninth were flagged, that performance would merit a score of -1. The critical value for the data snooping method was 3.29. The critical value for the tau test was 2.89. For the robust estimation method and the L1 norm minimization, an observation was flagged if its residual exceeded four times its a priori standard deviation.

RESULTS

The final results for single perturbed observations are shown in Table 1. In this part of the experiment the highest score was achieved by the L1 norm. This method produced perfect scores where the blunders were 10 standard deviations or larger. The next two highest scores are by the robust estimation and tau test methods. Both effectively singled out outliers after 5σ blunders were introduced.

The results of the scoring system were similar when pairs of outlying observations were introduced into the network. This is shown in Table 2. A perfect score of 650 was achieved only by the L1 norm technique. Robust estimation scored fairly well but was plagued by numerical problems when blunders were large. In Tables 1 and 2, the methods listed down the left column are DS, TT, RE, and L1 for the four methods described above. Across the top are the blunder magnitudes 4σ, 5σ, 10σ, etc. The score for each case is shown in the matrix, and the scores are totaled in the last column. In some cases scores were not available because of numerical problems.

The tau statistic seems to perform better as the outlier becomes larger. It effectively isolates blunders without flagging unblundered observations. As the blunder magnitude becomes larger the data snooping method seems to flag too many observations which adversely affects its score.

Table 1 Single Perturbed Observations					
Method	4σ	5σ	10σ	20σ	50σ	TOTAL
DS	16	23	2	-70	-226	-255
TT	6	17	24	24	24	95
RE	11	19	26	26	0	82
L1	10	22	26	26	26	110

0 indicates score not available, perfect score = 26.

Table 2 Pairs of Perturbed Observations					
Method	4σ	5σ	10σ	20σ	50σ	TOTAL
DS	386	493	-4	-1321	-3719	-4165
TT	28	55	252	365	394	1094
RE	248	474	644	0	0	1366
L1	259	535	650	650	650	2744

0 indicates score not available, perfect score = 650.

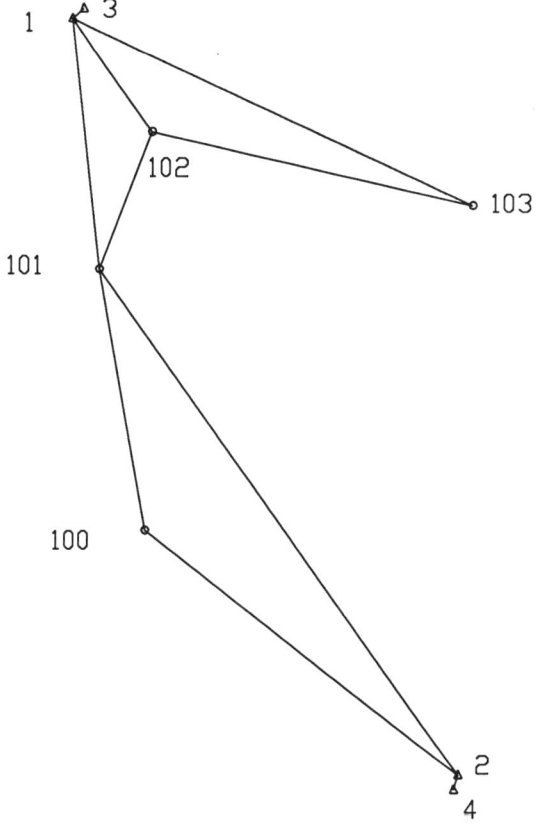

Figure 1. Test Network

CONCLUSIONS

The minimization of the L1 norm seemed to be the most effective method of detecting and isolating blunders. Further investigation yielded positive results for observations perturbed by over 10000σ. This would greatly aid in discovering blunders such as transposed digits or large systematic errors. This technique would isolate up to six blunders into the test network at once and still had a near perfect score when seven blunders were introduced. This would seem to be a good method for eliminating large blunders in a data set before conventional least squares adjustment is used. In this sense it is a pre-adjustment technique rather than a post-adjustment technique as are the others. For large networks, one would have to find a way to exploit sparseness in the matrices. This was not done for the small problem considered here.

Any of the first three techniques could be used after the fourth. Of these, either robust estimation or the tau test seems to be most effective. The former gives interpretable values for the residuals while the latter has a statistically determined critical value. However this comes at the expense of computing Qvv.

The test data used to conduct this investigation had a redundancy of 16 which is large. Not all practicing surveyors and geodesists can afford this luxury. The next step would be to conduct these same tests using a network with much smaller redundancy and compare those results to the ones presented here.

REFERENCES

Barrodale and Young, Algorithms for Best L1 and L∞ Linear Approximations on a Discrete Set, Numerische Mathematik 8, pp 295-306, 1966.

Fuchs, Interactive Bundle Adjustment with Metric and Non-Metric Images Including Terrestrial Observations and Conditions, ISP Commission III, vol. XXV, part A3a, 1984.

Kavouris, On the Detection of Outliers and the Determination of Reliability in Geodetic Networks, MS Thesis, University of New Brunswick, 1982.

Krarup, Goetterdammerung over Least Squares Adjustment, 44th ISP Congress, Commission III, 1980.

Mikhail, Review and Some Thoughts on the Assessment of Aerial Triangulation Results. Presented at Aerial Triangulation Symposium, Dept. of Surveying, University of Queensland, October 1979.

Pope, The Statistics of Residuals and Outliers, NOAA Technical Report NOS 65 NGS 1, Geodetic Research and Development Laboratory, NGS, Rockville, MD, 1976.

DESIGN OF A LOCAL COORDINATE SYSTEM
FOR SURVEYING, ENGINEERING, & LIS/GIS

Earl F. Burkholder, PLS, PE
Geodetic Engineer
Oregon Institute of Technology
Klamath Falls, Oregon 97601

BIOGRAPHICAL SKETCH

Earl F. Burkholder has taught upper division surveying classes at the Oregon Institute of Technology since 1980. He is on sabbatical leave during the 1990-91 academic year and studying at the University of Maine to learn more about surveying education and applications of new technology to modern survey practice. A graduate of the University of Michigan (BSCE 1973) and Purdue University (MSCE 1980), he worked with state plane coordinate systems and map projections on numerous engineering projects while employed by Commonwealth Associates Inc from 1973 to 1978. During the summer of 1983 he worked at the National Geodetic Survey researching material for NOAA Manual NOS NGS 5, The State Plane Coordinate System of 1983. He was Editor of the ASCE Journal of Surveying Engineering from 1985 to 1989.

ABSTRACT

State plane coordinate systems were designed to permit surveyors, engineers, and others to work with plane rectangular coordinates while enjoying the benefits of using the precisely surveyed National Geodetic Reference System (NGRS). The state plane systems are functional and useful, but lack complete acceptance because: 1) Benefits are deemed not worth the extra data collection and computational effort. 2) People avoid using what they don't understood. And 3), grid distances differ from the horizontal ground distance. These obstacles are surmountable through enforcement and education but, with computer processing resources so readily available, another solution has become increasingly attractive. By designing and implementing a local (county) coordinate system, benefits of using the NGRS and existing state plane coordinates can be retained, grid and ground distances are very nearly identical, and users continue to work with convenient rectangular coordinates. This paper is written to support ideas proposed by Nancy von Meyer in an article on "County Coordinate Systems" published in the June 1990 ACSM Bulletin. It goes on to discuss specific criteria necessary to preserve geometrical integrity of the NGRS and data sharing between data bases.

AUTHOR'S NOTE

I intended to meet the deadline for this paper to be included in the Proceedings for this Annual Meeting. The total scope of the paper was well defined when the proposal for same was submitted. However, it became obvious during discussions with surveying faculty at the University of Maine that certain specific recommendations were premature. Therefore, this Final Abstract is offered in lieu of the completed paper which will be distributed to attendees of the session and available upon request following the Annual Meeting.

Accurate Height Transfer from Bench Marks to Awkward Sites using GPS

Albert K Chong

Department of Surveying
University of Otago
Dunedin, New Zealand

Abstract

Elevations can be accurately determined by GPS techniques. Vertical control can be established on a routine basis without needing to use precise ephemeris and accurate geoidal models. This article demonstrates the suitability of GPS to establish elevations at awkward sites.

An awkward site is selected to illustrate the efficiency and simplicity of the GPS method for elevation determination. Some examples of awkward sites are roofs of high rise buildings, canyons, banks of rivers and dam structures. Various techniques are used to show the determination of orthometric height from ellipsoidal height, namely: local geoidal undulation difference by interpolation; scale plus rotations; and, estimated geoidal undulation.

Introduction

Accurate elevation is normally established by conventional surveying methods, namely: spirit levelling; trigonometric levelling; and, stadia levelling. A moderately accurate photogrammetric method is also applied occasionally. Recently, the introduction of Global Positioning System (GPS) has assured a new method of elevation determination. GPS technology is now widely recognised by surveyors as the most advanced surveying tool. It is used to obtain vertical control for photogrammetry; it is applied in earth and structure deformation to study vertical movement; and, it has been proven successful for many other types of surveying needs (Hajela, D., 1990; Collins, J., 1989).

Elevations are often required at awkward sites. A frequent occurrence is in the establishment of photogrammetric vertical control. Other applications include the establishment of elevation controls on dam structures and high rise buildings. The latter was selected for discussion in this paper. Although the need for accurate height control on high rise building is seldom essential, using conventional surveying method to establish control is always a time consuming, laborious and costly exercise (Bordley, 1989). Spirit levelling for this purpose is often infeasible. Trigonometric levelling is only suitable for a limited building height. For a very tall building the method becomes difficult and accuracy drops significantly.

The simplicity of GPS field operation is well documented (eg. Bordley, 1989). GPS receivers are simply set up on survey tripods, levelled and centred over bench marks. Antenna heights are measured and the receivers are left to collect data. Records of field atmospheric condition are usually not essential and GPS measurements are seldom affected by poor weather.

Ellipsoidal Height and Orthometric Height

GPS is a three dimensional measurement system. Observations are processed to give a survey mark position in cartesian coordinates (X, Y, Z), which can be converted easily to geodetic coordinates, i.e. latitude, longitude and height-above reference ellipsoid. The computed height-above reference ellipsoid can be converted to orthometric height or elevation for many applications. Observations from relative GPS surveying is most useful because accurate elevation can be determined mathematically for points of unknown elevation by connecting the GPS traverse to one or more existing bench marks. A few mathematical techniques have been developed to carry out the task. To select a technique it is necessary to examine the accuracy needed for a project, the GPS observations available and the type of existing control occupied. For example, height difference based on height-above-reference ellipsoid is adequate for earth/structural deformation study while height difference based on height-above geoid is essential for many civil engineering projects. The relationship between the ellipsoidal height and orthometric height is shown in numerous published articles (Collins, 1989; and, Kearsley, 1988). Three functional mathematical techniques are discussed below.

Direct Deduction

This technique, consisting of two approaches is suitable in many situations, particularly for a site where there is none or only one bench mark available. Detail knowledge of the geoid-ellipsoid separation (N) in the local area will yield accurate orthometric height. Estimated (N) can produce good orthometric height for many civil applications. The computations required for the two approaches are slightly different and are described separately below.

a) **Ellipsoid height difference**

One simply adds the GPS ellipsoid height difference between Bench Mark (A) and new point (B) to the elevation of (A) to obtain the elevation of (B).

    For example:

        Given bench mark (2001*) elevation is 13.316 m
        Ellipsoid height difference (2001-3001) is +18.392 m
        Elevation of Point 3001 = 31.708 m

To compute the elevation of points in a GPS network, a minimally constrained adjustment should be performed to eliminate any blunder and error in the network.

b) **Geoid undulation difference**

The result in (a) can be improved by using a known or an estimated geoid undulation difference. In the case where a bench mark is not available and an estimated orthometric height is required then this method should be used. Extensive physical gravity observations are essential to determine local geoid-ellipsoid separation. The task of gathering physical gravity observations can be expensive. However, estimation can be performed by using a computer program known as GEOID360F, which was based on a geopotential model using spherical harmonics. The model was developed by Professor Richard Rapp at the Ohio State University. A PC based version of the computer program is now readily available to GPS user. The result of the example in (a) above is further refined by adding the estimated geoid undulation difference, i.e.

    Predicted geoid height at 2001 is 3.721 m
    Predicted geoid height at 3001 is 3.701 m
    Estimated undulation difference (2001-3001) = -.020 m
    Elevation of Point 3001 now = 31.688 m.

Fixing One Elevation

This technique can be applied when two or more bench marks are tied to the GPS network. After the network is adjusted by a minimally constrained least square method, the elevation of a centrally located bench mark is fixed in a constrained network adjustment. This method is suitable for a project where the site covers a small area or a narrow strip where geoid height varies substantially. The residuals in elevation of other non-fixed bench marks are used in a graphical smoothing as depicted in figure 1. Residual correction is applied to the elevation of all new points in the network.

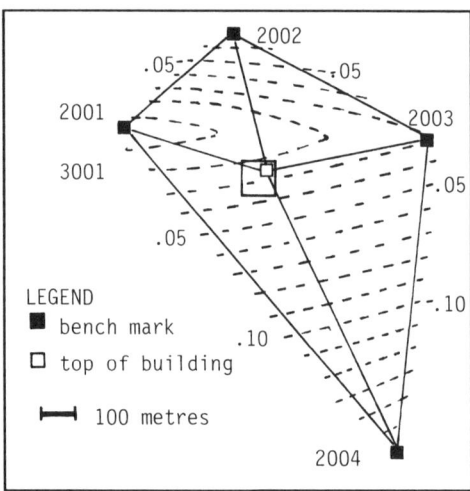

Figure 1 Corrections for Geoid Undulation

Scale and Rotations

This is a frequently used technique (FGCC, 1989, p30). It is basically a constrained three-dimensional adjustment in which either an average scale is computed or scale plus rotations are computed. In fact both sets of adjustment should be performed in any GPS network where three or more bench marks are available. If the standard error of any of the three computed rotations is large, then an average scale adjustment or a 'fixing one elevation' adjustment should be evaluated.

Tables 1a and 1b, show the size of residuals of average scale and the scale plus rotations adjustment respectively. The standard error of the average scale is rather small in comparison with those of the scale plus rotations. The size of the standard error gives an indication of the reliability of the two adjustments.

PARAMETER	ADJ. VALUE	ST. ERROR
Average Scale	28.201	1.180

1a

PARAMETER	ADJ. VALUE	St. Error
Scale	10.5082	48.7461
Rotation X	0 0 08.41	0 0 39.22
Rotation Y	0 0 13.53	0 0 24.51
Rotation Z	0 0 36.19	0 0 39.10

1b

Table 1 Parameters and Their Standard Errors of
a) Average Scale and b) Scale Plus Rotations

Awkward Project Site

Accurate height of the top of a concrete pillar on the roof of a seven-story university building is needed for a surveying project. To test the capability of the relative static GPS surveying technique, GPS receivers were hired to carry out the survey. An elevation of the pillar was previously obtained by the trigonometric levelling method. The top of the pillar can be seen a kilometre away from the building. A number of first and second order bench marks are conveniently located near the building.

Four bench marks were selected for the GPS project. The pillar was connected to each bench mark by GPS base lines. Furthermore a close traverse loop connecting all the bench marks was also observed by GPS.

Results of Survey

GPS observations were post-processed and analyzed according to the techniques discussed previously in this article. All the GPS-measured base lines of the project were accepted for the final analysis. The results of the analysis are tabulated in Table 2. The elevation of the surveyed pillar obtained by the trigonometric method is provided at the bottom of the table. Table 3 shows the difference in elevation of the various techniques using the trigonometric levelling result as the norm.

ELEVATION ANALYSIS TECHNIQUE	BM USED	ELEVATION OF PILLAR (3001) in Metre
1) Direct Deduction a) Ellipsoidal height difference b) Geoidal undulation difference	 2001 2001	 31.708 31.688
2) Fixing One Elevation	2001	31.672
3) Scale and Rotations (both sets use the same BM) a) Average scale b) Scale plus rotations	 2001 2002 2003 2004	 31.676 31.716
4) Trigonometric Levelling		31.664±5 mm

Table 2 Elevation of Pillar by Various Techniques

ELEVATION ANALYSIS TECHNIQUES	DIFFERENCE IN ELEVATION (using trigonometric levelling result as the norm) in Metre
1) Direct Deduction a) Ellipsoidal height difference b) Geoidal undulation difference	 .043 .023
2) Fixing One Elevation	.007
3) Scale and Rotations a) Average scale b) Scale plus rotations	 .011 .051
4) Trigonometric Levelling	.000 ±5 mm

Table 3 Elevation Difference Between Various Techniques

Conclusions

As anticipated the result obtained by 'fixing one elevation' is closest to the result of the trigonometric levelling survey. The expectation is based on the fact that the geoid undulation varies by .020 m as determined by the GEOID360F program. Thus an interpolation of the height difference in the locality is appropriate.

The average scale technique gave a better elevation than the scale plus rotations technique. Therefore it is worthwhile to note that scale plus rotations should not be used indiscriminately without checking out other techniques.

The elevation obtained by GPS survey was within a few millimetres of the value acquired by trigonometric method. The high accuracy of GPS and the simplicity of its field observation are making the technology suitable for height transfer to awkward locations. For subsidence and deformation study of most high rise buildings it may not be necessary to convert ellipsoidal height difference to orthometric height difference, thus simplify the data reduction process even further.

References

Bordley, R., 1989. Taking the Spirit Out of Levelling, *L&MS*, UK, Vol. 7 (11), p526-534.

Collins, J. 1989. Fundamental of GPS Baseline and Height Determinations, *Journal of Surveying Engineering, ASCE*, USA, Vol. 115 (2), p223-238.

Federal Geodetic Control Committee (FGCC), 1989. *Geometric Geodetic Accuracy Standards and Specification for Using GPS Relative Positioning Techniques*, Version 5, *NOAA* MD USA 48p.

Hajela, D., 1990. Obtaining Centimetre-Precision Heights By GPS Observations over Small Areas. *GPS WORLD*, Oregon USA, Vol.1 (1), p55-59.

Kearsley, A.H.W, 1988. The Determination of the Geoid Ellipsoid Separation for GPS Levelling, *The Australian Surveyor*, Vol.34 (1), p11-18.

USE OF CONSTITUENT ANALYSIS FOR ESTIMATION OF TIDAL DATA
BY SIMULTANEOUS SHORT-TERM OBSERVATIONS

George M. Cole and F. Michael Speed

BIOGRAPHICAL SKETCHES

George M. Cole is a registered engineer and surveyor currently serving as President of Florida Engineering Services Corporation of Tallahassee, Florida. His professional background includes service as the State Cadastral Surveyor and as coordinator of the coastal mapping program for Florida, and service with the U.S. Coast and Geodetic Survey. He has a lengthy record of research and publications in the field of water boundaries.

Dr. F. Michael Speed currently holds the Blucher Chair for Surveying and Science at Corpus Christi State University. His background includes assignments as Director of the Blucher Institute at Corpus Christi State; a scientist for the NASA Manned Spacecraft Center; an instructor in mathematics, statistics and computer sciences at universities in Texas, Louisiana and Mississippi; and as a partner in an oil/gas production company. Dr. Speed has a distinguished record of research, consultation and publication in the statistics area.

ABSTRACT

Although a tidal datum is usually defined as an average value over a complete tidal epoch of 19 years, most datum determinations use observations of considerably shorter duration. Average values from such short-term observations are corrected to the equivalent of a 19-year mean by a process of simultaneous comparison with observations taken at a control tide station with a known 19-year mean datum. This paper proposes an alternate correlation process which uses observed tidal data from throughout the tidal cycles within the observational period, not just the high and low extremes. The proposed process allows significantly less frequently sampled data, eliminates the need for selecting the high and low tidal extremes, and allows the use of non-continuous and random length observational periods. Tests of the method with both theoretical and observed data are included.

INTRODUCTION

A tidal datum is usually defined as an average of a certain tidal extreme over a long-term period. Generally, a period of 19 years is used to include all of the significant constituent cycles composing a typical tide. Such constituents range in period from 14 minutes to 18.6 years. For economics, tidal data at most locations are determined by observations over periods of considerably less than 19 years. Such short-term observations are taken simultaneously with observations at a control station, where 19-year mean values have been determined, and reduced to the equivalent of a 19-year mean by a correlation process.

CURRENT CORRELATION PROCEDURES

Historically, this correlation has been performed by a relatively unsophisticated process (Marmer, 1951) which compares the observed short-term mean tide level and mean tidal range with the known 19-year mean tide level and mean tidal range at the control station. Corrections resulting from this comparison are then applied to the observed short-term mean tide level and mean tide range at the subordinate station. Only the high and low extremes of tidal cycles are used for this process.

These procedures may be expressed algebraically by the following formulae (Cole, 1983):

Notation
- MHW = 19-year mean high water
- MTL = 19-year mean (half) tide level
- MLW = 19-year mean low water
- MR = 19-year mean range
- TL = mean half tide level for observational period
- R = mean range for observational period
- s = subscript used to denote subordinate station
- c = subscript used to denote control station

The first formula calculates the subordinate 19-year mean range:

$$MR_s/R_s = MR_c/R_c \qquad (1)$$

This may be restated as follows:

$$MR_s = R_s/R_c \times MR_c \qquad (1a)$$

The second formula calculates the subordinate 19-year mean tide level:

$$TL_c - MTL_c = TL_s - MTL_s \qquad (2)$$

This may be restated as follows:

$$MTL_s = (TL_s - TL_c) + MTL_c \qquad (2a)$$

Formulae (3) and (4) calculate the 19-year mean high and mean low water by applying half of the mean range to the mean tide level.

$$MHW_s = MTL_s + MR_s/2 \qquad (3)$$

$$MLW_s = MTL_s - MR_s/2 \qquad (4)$$

From examination of equations (1a) and (2a) above, it may be readily seen that the 19-year values at the subordinate station are defined as a function of the 19-year mean value at the control station and the relationship between the observed values at the control and subordinate stations.

LIMITATIONS OF CURRENT PROCEDURES

The current procedures are a product of earlier limitations of data acquisition in that they use only the tidal extremes (high and low tides). "Prior to the invention of the automatic tide gauge, the recording of tides throughout the 24 hours of the day was a matter of considerable expense. It was therefore customary to observe the tide only near the time of high and low water. This permitted a tabulation of the high and low water, but not of the hourly heights. Half tide level could be determined from such tabulations, but not mean tide level." (Marmer, 1951) Today's recording gauges measure tides throughout the tidal cycle. Typically, ten observations per hour are taken to insure capture of the absolute tidal extremes, resulting in massive amounts of data collected. Use of only two points out of a complete tidal cycle for correlation is wasteful of data and not as statistically valid as using all observed data points. Secondary problems are created by this process in accurately selecting the "true" tidal extremes. These include not only the question of data smoothing but also problems in areas where meteorological effects mask tidal peaks.

In addition, the standard method is limited by its assumption that a linear relationship exists between the two tide stations. This is evidenced by equations 1 (a) and 2 (a) which take the form of the equation of a straight line ($Y = a + bX$). This assumption is valid only when the period of observation is equal to an integral number of the periods of all of the constituents for which there is an amplitude or phase difference between the two stations. The standard method compensates for this limitation by requiring the averaging of simultaneous observations for all of the tidal cycles over a period corresponding to the longest period constituent for which the data analysist believes there may be a phase or amplitude difference between the two stations. Furthermore, this concept requires that the data be continuous during that period at both stations for statistical validity. This requirement can result in the waste of many months of data due to a few days of tide gauge malfunction at one of the stations during an observational period.

PROPOSED APPROACH

The problem in accurately correlating tidal observations between two stations becomes evident when it is recognized that a tide "wave" is really a compound wave, a composite of numerous constituent waves. Both the phase and amplitude relationship for each constituent wave at the stations should be used. Therefore, it is proposed that such an approach be used for the correlation.

The individual constituent waves can not be measured directly, since they are masked within the composite tidal wave. However, they may be derived from observational data by the statistical process of harmonic analysis. The proposed approach would therefore use harmonic analysis to derive the phase lag and amplitude of the tidal constituents from the observed data at the control and subordinate station. Periodic observations from throughout the observed tidal cycles would be used for this analysis, as opposed to only the tidal peaks. It is noted that values for constituents with periods significantly longer than the observational period may not be determined by the harmonic analysis. Therefore, as in the standard method, the proposed method requires observations, although not necessarily continuous observations as in the standard method, over substantially all of any period for which there are differences in constituent values between the two stations. However, sampling during that period may be considerably less frequent than required by current procedures. For example, hourly sampling is adequate as opposed to the tenth of an hour sampling rate currently used.

Once the amplitudes of the constituents are determined by harmonic analysis, the composite tidal range for both the control and the subordinate station may be approximated by the following formula (Derived from Coast & Geodetic Survey, 1952, p 9):

$$R = 2 A_m + 0.5 A_m \sum_{\substack{i=1 \\ i \neq m}}^{37} (A_i S_i)^2 / (A_m S_m)^2 \quad (5)$$

Where R = Composite tidal range
       A = Amplitude of constituents
       S = Speed of constituents in Deg/Hr
       i = Constituent number (Using NOS constituent list
       m = Constituent number of primary constituent (M2 constituent for semi-diurnal or O1 for diurnal tides)

Since the observed amplitudes of the constituents may vary from time to time, as well as due to the lack of knowledge of the longer term constituents, the ranges derived from equation (5) may vary somewhat from the 19-year mean range. Therefore, the ratio of the two composite tidal ranges derived from this process are used in formula 1(a), of the standard method, to calculate the 19-year mean range at the subordinate station (MRs = Rs/Rc x MRc).

The determination of 19-year mean tide level by the proposed method is somewhat more complex, although easily handled by computer. The proposed method involves determination of mean sea level for the observational period at the two stations by use of the constituent values derived from the harmonic analysis together with the actual water level observations. Then, the observed half tide levels are determined from the mean sea level values and are used to calculate 19-year mean tide level by use of equation 2(a) of the standard method (MTLs = TLs - TLc + MTLc).

To derive observed mean sea level at the two stations for use in this step, tidal predictions are made for the period of observations. These are made using the following equation (Derived from Schureman, 1958):

$$H = \sum_{i=1}^{37} A_i \cos(S_i T - K_i) \qquad (6)$$

Where H = Height of tide at time T
K_i = Phase lag of constituent i in degrees.
A_i = Mean amplitude of constituent i
T = Time reckoned from beginning of year in hours
S_i = Speed of constituent i in degrees per hour
i = Constituent number (from NOS constituent list)

Predictions using this formula should approximate the observed tide less mean sea level. For this process, predictions are made for the time of each observation of tidal height. This predicted height is then compared with the observed height. The mean of the residual differences between observed and predicted values would then approximate mean sea level for the observational period. Observed mean half tide level may then be determined by the following formula (Marmer, 1951):

$$TL = SL + A_4 \cos(2K_1 - K_5) - \frac{0.03(A_4 + A_6)^2}{A_1} \times \cos(K_1 - K_4 - K_6) \qquad (7)$$

A summary of the steps involved in the proposed process are as follows:

　　1) Perform harmonic analysis on simultaneously observed tides at the control and subordinate stations to derive phase lag and amplitude for each constituent.

　　2) Determine approximate composite tidal range for the control and subordinate stations using Equation 5.

　　3) Determine 19-year mean range for subordinate station by Equation 1(a) using the ratio of tidal ranges, derived in Step 2, for Rs/Rc.

4) Predict tidal heights, using Equation 6, for the times of observation at both stations.

5) Determine approximate sea level for both stations by obtaining mean of differences between observed tidal height and predicted tidal height for all observations.

6) Determine approximate half tide level for both stations using equation 7.

7) Determine 19-year mean tide level at the subordinate station using Equation 2(a).

TESTS OF PROPOSED METHOD WITH THEORETICAL DATA

To evaluate the proposed approach, tests using theoretical data have been performed. Use of such data allows a more controlled test together with the ability to freely manipulate various parameters. For these tests, a model was used to generate a simplified four constituent tide with an arbitrary mean sea level reading of 13.00 feet and with deliberately widely differing constituent values as follows:

TABLE 1

Symbol	Period	Speed Deg/Hr	Amplitude(Ft) Control	Amplitude(Ft) Subord.	Phase Lag(Deg) Control	Phase Lag(Deg) Subord.
M2	12.42 Hrs	28.9841042	3.0	2.0	0	45
S2	12.00 Hrs	30.0000000	0.6	0.4	0	0
Mm	27.57 Days	0.5443747	1.0	1.0	0	180
Sa	1 Yr	0.0410686	2.0	2.0	0	0

Primary Determination

Since the longest period constituent comprising the model tide is one year, a primary datum determination for the test tide stations can be accomplished in one year. This was accomplished using peaks created by a prediction using equation (6). This resulted in the following long-term values for the model tides which may be considered the "correct" values for evaluating the results of secondary determinations:

	CONTROL	SUBORDINATE
MHW	16.04	15.00
MTL	13.01	12.98
MLW	9.98	10.97
MR	6.06	4.03

Results Using Standard Method

Since there are no differences in constituent values for the long period constituent (Sa), a 28 day period should suffice, under the constraints of the standard method, to determine a long-term datum at the subordinate station. The standard method (Equations 1 through 4) was then used to estimate tidal data for the subordinate station using observations for the complete 28 day period and then repeated with an assumption that observations were not available for the portion of the 28 day period from day 11 through day 17. The results of these calculations are illustrated in Table 2.

As may be seen, using all data, the agreement is fairly good between the estimated data and data from the primary determination. A difference of 0.05 feet is noted in mean range and 0.03 feet in mean tide level. When using the non-continuous data, however, a difference of 0.13 feet in mean range and 0.57 feet in mean tide level exists between the estimated values and those resulting from the primary determination. As expected, the accuracy of the results degenerated due to using other than the complete 28 day period when using the standard method.

Results Using Proposed Method

Using the same non-continuous data set but with only hourly readings, an analysis was performed using the proposed method with the results also illustrated in Table II.

TABLE 2

PRIMARY DETERMINATION		ESTIMATION WITH SHORT TERM DATA		
		STANDARD 28 days	STANDARD Missing Data	PROPOSED Missing Data
MT_{Lc}	13.01	T_{Lc} 14.92	15.22	13.00
MR_c	6.06	R_c 5.97	5.66	6.06
		T_{Ls} 14.92	14.62	13.00
		R_s 3.89	3.89	4.04
MT_{Ls}	12.98	13.01	12.41	13.01
MR_s	4.03	4.08	4.16	4.04

As may be seen, with theoretical data, the proposed method produced results with excellent agreement with the primary determination, even while using only hourly readings and with gaps in the data which caused inaccuracies using the standard method. As a further demonstration of the potential of this method, the process was repeated with only two observations per day and with no observations during the seven "missing" days. Therefore only 40 observations scattered throughout the 28 day period were used. Identical results as those obtained using all of the hourly readings were obtained.

TESTS OF PROPOSED METHOD WITH REAL DATA

In addition to tests with model tides, tests were also conducted with actual observed tidal data. For the first test, data from the National Ocean Service's tide stations at Fernandina Beach and Mayport, Florida were used. Observations have been conducted simultaneously at both stations for at least 19 years. This allows for primary determinations of tidal data for both stations for evaluation of the results of the secondary determinations. Using observed data for two years chosen at random, an estimate was made for Mayport using both the standard and proposed methods for both years. For input into the standard method, annual mean data computed by the National Ocean Service were used. For input into the proposed method, hourly heights from the observations over the selected year were used.

In the second test, only two water level observations per day were used for both stations used in test one. The standard method could not be used with such limited data. As may be seen in Table 3, results similiar to those obtained in test one resulted with the proposed method.

The third test is a worst case scenerio. Two stations were selected from areas with limited tidal range, where the long term constituents are greater than the daily constituents, and where meteorlogical effects tend to mask the tidal effects. Such areas present technical as well as legal complexities when attempting to use tidal data for coastal boundaries since the predominate water cycle is other than that caused by the daily or semi-daily tide. As may be seen from Table 3, the proposed method tended to break down under such conditions.

TABLE 3

Primary Determ. Results	Year	Estimation With Short-Term Data			
		Standard		Proposed	
		Input	Results	Input	Results

Test 1. Fernandina Beach vs. Mayport, Florida (1971-1989)

MHWs= 6.26	86	MTLc=4.75	MHWs= 6.23	Rc =5.94	MHWs = 6.25
MTLs= 4.01		MRc =6.04	MTLs= 3.97	Rs =4.52	MTLs = 3.95
MLWs= 1.76		Rc =5.87	MLWs= 1.71	SLc =5.00	MLWs = 1.65
MRs = 4.50		Rs =4.39	MRs = 4.52	SLs =4.16	MRs = 4.60
		TLc =4.90		TLc =4.93	
		TLs =4.12		TLs =4.13	
MHWs= 6.26	87	MTLc=4.75	MHWs= 6.22	Rc =5.89	MHWs = 6.27
MTLs= 4.01		MRc =6.04	MTLs= 3.96	Rs =4.48	MTLs = 3.97
MLWs= 1.76		Rc =5.85	MLWs= 1.70	SLc =4.93	MLWs = 1.67
MRs = 4.50		Rs =4.37	MRs = 4.51	SLs =4.12	MRs = 4.59
		TLc =4.84		TLc =4.87	
		TLs =4.06		TLs =4.09	

Test 2. Fernandina Beach vs. Mayport, Florida (1971-1989)

MHWs= 6.26   86 (Standard Method can   Rc =5.94   MHWs =6.25
MTLs= 4.01   not be used with this   Rs =4.52   MTLs =3.95
MLWs= 1.76   data set)   SLc =5.00   MLWs =1.65
MRs = 4.50   SLs =4.16   MRs =4.60
                                        TLc =4.93
                                        TLs =4.13

Test 3. Port Isabel vs. Port Mansfield, Texas
(1965-1983)

MHWs= 2.57 82 MTLc=4.36 MHWs= 2.54   Rc =1.04   MHWs= 2.31
MTLs= 2.43   MRc =1.38 MTLs= 2.40   Rs = 0.06 MTLs= 2.26
MLWs= 2.29   Rc =1.33 MLWs= 2.26   SLc = 4.46 MLWs= 2.16
MRs = 0.28   Rs =0.26 MRs = 0.27   SLs = 2.36 MRs = 0.10
                TLc =4.34   TLc = 4.46
                TLs =2.38   TLs = 2.36

It is noted that, for Port Isabel and Port Mansfield, data for mean higher high water and diurnal range were used for MHW and MR.

SUMMARY

A proposed new approach has been suggested for correlation between tide stations for secondary determinations of long-term tidal data. This approach allows a significant reduction in the frequency of sampling needed and eliminates the need for selecting high and low tidal extremes. In addition, the proposed method allows the use of non-continuous data, and allows the use of data from random length observational periods.

The results of tests of the proposed method using theoretical data were presented to illustrate the approach and its advantages. These tests demonstrate that excellent agreement, superior to the standard method, with data derived from primary determinations is possible even with significant gaps in the observed data which causes the standard method to produce incorrect results and with a significant reduction in the amount of data required. Results of tests with observed data were also presented. Those data indicated that the proposed method produced acceptable results in areas with "conventional" tidal patterns with great reductions in data requirements. However, less than accurate results were obtained in areas where long-term constituents were greater than daily constituents and where tidal variations were significantly masked by non-tidal water level changes, perhaps confirming the previously mentioned non-applicability of conventional tidal data for boundaries and/or sounding data in such areas.

In conventional tidal areas, however, the results from these preliminary tests are encouraging and suggest that refinement of this method will result in a production procedure superior to current procedures.

LITERATURE CITATIONS

Coast & Geodetic Survey, Manual of Harmonic Constant Reductions, Special Publication 260, Govenment Printing Office, 1952

Cole, George M., Water Boundaries, Landmark Publications, Rancho Cordova, CA, 1983

Marmer, H.A., Tidal Datum Planes, Government Printing Office, 1951

Schureman, Paul, Manual of Harmonic Analysis and Prediction of Tides, Special Publication No. 98, Government Printing Office, 1940

GPS PROJECT DESIGN: THE OWENS VALLEY EXPERIENCE

James K. Crossfield
Wesley Parks
Barbara S. Littell
Department of Civil and Surveying Engineering
California State University, Fresno
Fresno, California 93740-0094

BIOGRAPHICAL SKETCHES

Professor James K. Crossfield, chairs the Northern California Section of ACSM and coordinates the CSU, Fresno Continuing Engineering Program. He also serves as faculty adviser for the CSUF Surveying and Photogrammetry Student Association, The CSUF ACSM Student Chapter, and the ForeSight! Newsletter.

Dr. Wesley Parks is a Surveying Engineering graduate student at CSU, Fresno. He has had extensive work experience with the U.S. Fisheries Service. As a research assistant, Wes played a key data analysis role for the Owens Valley GPS Project.

Barbara S. Littell is Vice President of the CSU, Fresno ACSM Student Chapter. She has contributed articles to the ForeSight! Newsletter and is currently a CO-OP student with NGS. As a research assistant, Barbara was instrumental in developing detailed station reconnaisance information for the Owens Valley GPS Project.

ABSTRACT

An optimized GPS observational scheme was developed for the Los Angeles Department of Water and Power Owens Valley-Mono Basin GPS Control Survey. Approximately 175 planned independent baselines connected 90 network points along a twenty mile wide, 250 mile long corridor. The design, developed according to First Order FGCC criteria, was organized into daily (one and two session) observational periods. Example daily sketch maps and transportation plans are illustrated. Operational considerations included helicopter transportation to remote mountain peaks, up to two hour travel times between sessions, coordination of six dual-frequency GPS receivers and plan modifications necessitated by actual field conditions. Procedures for creating optimized GPS network designs are discussed.

INTRODUCTION

This discussion provides a detailed look at the GPS network design provided to the Los Angeles Department of Water and Power (LAWP) for the Summer 1990 observational campaign. Comprehensive First-Order (F.G.C.C., 1988) area-wide horizontal and vertical geodetic survey control was needed throughout the Owens Valley-Mono Basin to support the creation of a system-wide Geographic Information System.

PRELIMINARY PREPARATION

Detailed network design and observational plans cannot be developed without the benefit of a complete field reconnaisance. This reconnaisance insures that proposed network control points will be usable when needed. A detailed field reconnaisance was conducted for this project during a three week period during the later part of June 1990.

All proposed network points were checked for skyward obstructions, and had any existing obstructions carefully plotted so that observational plans could be developed around them. Station location sketches and how to reach descriptions were developed. A comprehensive project map was created, showing the proposed locations of all network points. Color and symbol coding was utilized to help differentiate the various types of points (horizontal control, vertical control, new point, etc.) on the project map. Station sketches, how to reach descriptions, preliminary coordinate data, station locality photographs, sky charts and associated reconnaisance notes were consolidated for each proposed network station into a series of three ring binders.

Finally, after consultation with LAWP officials, 90 points were selected for inclusion in the network. All extraneous points were removed from the project map, and a tracing paper overlay was created, with all network points plotted at the same scale (1:250,000) as the project map.

NETWORK DESIGN

The network was designed to meet or exceed First Order FGCC criteria. Pencil lines were drawn on the overlay map representing proposed baselines so as to conform to the following generalized criteria.

1. No primary loop contained more than six independent baselines.
2. Each loop contained observations from at least two different sessions.
3. Two hour (minimum) dual-frequency observational sessions were planned.
4. Each network point had at least three independent baselines connected to it.

Several baselines could fail and the overall network adjustment would not be seriously affected. The proposed baseline observations were penciled onto the overlay map. Modifications were made as necessary to conform to the generalized criteria, and to insure a balanced array of independent baselines interconnecting the overall network. Approximately 175 independent baseline observations were planned for this network. About 5 lines were planned duplicates. A large number of dependent baselines were generated (about 350). These dependent baselines help to isolate and detect observational errors and may in a few instances provide substitute baselines if a some of the planned independent baselines fail.

Table 1 provides a summary of the numbers and types of points included in the Owens Valley-Mono Lake GPS Control Survey. All primary control points, and several of the new points have published NAD 83 coordinates. Thirteen of the network stations have published elevation data (based on NAD 83) provided to the nearest centimeter.

Network Point Type Summary

Type Point	Number
Primary Control	30
Caltrans HPN	8
VLBI	2
NAD 83 Vertical	13
New LAWP Points	36

TABLE 1

Primary control points primarily consisted of two parallel lines of first order horizontal control points aligned along the Eastern crest of the Sierra Nevada Mountains and the Western crest of the White-Inyo Mountains. Most necessitated helicopter access. Only three could be considered "drive to". Caltrans High Precision Network (HPN) points were located along state and federal highways with about a 20-30 mile spacing. These points are planned for inclusion in the Caltrans statewide High Precision GPS Network. The NAD 83 vertical points were a mixed assortment of additional secondary horizontal control and various orders and classes of vertical control points throughout the project area that had geoid heights and elevations published to the nearest centimeter in NAD 83 coordinate booklets. A few of the other points selected also had "good" vertical data associated with them. The new LAWP points primarily consisted of existing second order horizontal control scattered across the project area. A few new points were established however in strategic locations.

Table 2 contains a description of the nomenclature used on the daily observation maps created for this project, an example of which is illustrated in Figure 1. Each of the six receivers was color coded to facilitate crew assignments and planning control. Each color code was designed to be permanently assigned to a field crew so that operational idiosyncracies could be monitored and controlled.

While any crew could operate any receiver (all receivers were the same), it was wise to maintain crew-color code discipline because the observational plan was designed to minimize the number of different points that each color coded receiver visited. Thus each crew needed to become familiar with fewer points. Receivers were assigned to network points according to optimization criteria developed in "Optimizing GPS Corridor Control Surveys" (Crossfield, 1990).

Network Design Nomenclature Summary

Symbol	Description
△ (with dot)	Primary Control point
△	Caltrans HPN Point
⬡ (with dot)	VLBI Point
◉	NAD 83 Vertical or new LAWP
-------	Session 1 Observation
- - - -	Session 2 Observation
G (11-2) W	Green Unit (Day 11 - Session 2) White Unit

TABLE 2

Planning constraints included the following issues:

1. The duration of elapsed helicopter support required minimization.
2. Two primary control stations (Coso and Volcano), located within the China Lake Navel Weapons Center, had to be observed before 1:00pm.
3. The field phase of the project had to be completed before August 21, 1990 as half of the project team was scheduled to depart on that date.
4. Transportation times, including initial setup and intra-session moves required minimization.
5. Certain primary control points needed to be visited only once if possible due to the difficult access by helicopter.
6. Due to the threat of afternoon thundershowers, the most exposed mountain peaks required occupation during the morning session.

Figure 1 represents the initial plan for day six. Six initial helicopter transports were required to position the crews on primary control points. Five inter-session moves by helicopter were required. This was the most travel intensive day associated with the entire project. The "Clear" color code crew for example, started the day at station Black and moved to station Laws for the second session. Use of the daily transportation plan, see Table 3, facilitated crew command and control for the project management staff. This table was correlated to a specific daily observational plan. Daily observational plans and transportation plans were prepared for the entire project before the field phase of the project began.

Several primary control points required only one visit. These were: Potato peak (Day 2), Mount Downs (Day 4), Chalfant (Day 6), Lookout (Day 9), Volcano and Wortley (Day 12), El Paso (Day 13), and Pajuela (Day 14). Thus 8 of the 27 helicopter only primary control points required only a single visit.

LAWP Owens Valley-Mono Basin GPS Survey
Day 6 Observation Plan

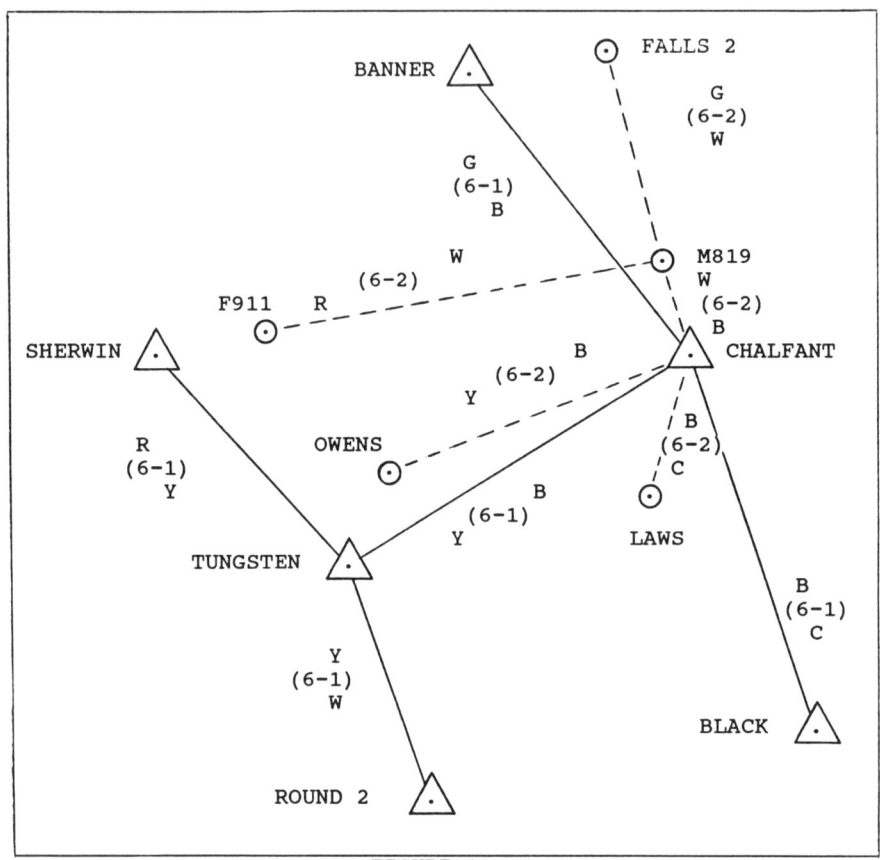

FIGURE 1
(Not to Scale)

The transportation plans were designed to minimize travel times and distances. Single person crews could be utilized on non remote points, particularly when no movement was involved. The full schedule was designed to initially allow 13 straight helicopter days. Helicopter usage was planned to be phased in during the first few days of the activity, followed by heavy use during the middle phases of the project when familiarity was assured. The last four field days were designed to need only ground vehicle transportation. These days could be utilized anytime if conditions at higher elevations became a problem. The project was planned to be conducted from north to south. Helicopter landing locations and support base camp locations became a major factor. A sequence of proposed helicopter base locations was provided in the initial planning guide.

LAWP Owens Valley-Mono Basin GPS Survey
DAILY TRANSPORTATION PLAN

DAY: 6 , DATE: Friday Aug. 3, 1990

Activity	Black	Red	White	Green	Yellow	Clear
Vehicles Initially Driven To	Bishop	F911	Bishop	Falls 2	Owens	Bishop
Initial Chopper Transports	To Chalfant	To Sherwin	To Round 2	To Banner	To Tungsten	To Black
Order	1	5	3	6	4	2
Chopper Moves	X	To F911	To M819	To Falls 2	To Owens	To Laws
Order	X	3	1	2	4	5
Vehicle Moves	X	X	X	X	X	X
Final Chopper Transports	To Bishop	X	To Bishop	X	X	To Bishop
Order	2	X	1	X	X	3

TABLE 3

LESSONS LEARNED

Actual operational considerations caused major revisions in the proposed day to day sequence, but each specific daily plan was retained virtually unchanged. A few modifications were easily handled late in the project to facilitate timely completion of all phases and to accommodate a successful operational test of the early evening session. The daily transportation plans were not as helpful as originally hoped. A modified daily transportation plan appears in Table 4. The major revision involved inclusion of proposed travel and move times, including times indicating when to arrive on the job. This is particularly critical when initial travel times fluctuate widely (as on this project) and when the data collection effort continues for a month or more causing satellite availability changes of hours.

ACKNOWLEDGEMENTS

The authors wish to thank Karl Riesen for being an effective catalyst and bringing CSU, Fresno Surveying Engineering expertise to the attention of Mr. Ronald McGhie and the entire LAWP Aqueduct Division in Bishop, California. CSU, Fresno Surveying Engineering students Karl Riesen, Gene Muse, Ed Patton and Bob Hagler, employed by LAWP during the project, are thanked for their tireless efforts to bring this project to a successful conclusion. The timely support provided by the surveys personal at the Bishop office of LAWP was greatly appreciated. Ron McGhie is especially thanked for providing us with this tremendous learning opportunity.

LAWP Owens Valley-Mono Basin GPS Survey
MODIFIED DAILY TRANSPORTATION PLAN

DAY: __6__, DATE: __Friday Aug. 3, 1990__

Crew Activity	Shereef Eric Black	Pat Red	Gene Barbara White	Gerry Ed Green	Wes Bob Yellow	Jim Karl Clear
Arrive Yd	0530	0630	0700	0630	0630	0530
Depart Yd	0600	0720	0700	0730	0700	0615
Vehicles Initially Driven To	Bishop	F911	Bishop	Falls 2	Owens	Bishop
Initial Chopper Transports	To Chalfant	To Sherwin	To Round 2	To Banner	To Tungsten	To Black
Time	0630	0820	0730	0845	0800	0700
Session 1 Time	0830-1200	0900-1200	0900-1200	0930-1200	0900-1200	0830-1200
Chopper Moves	X	To F911	To M819	To Falls 2	To Owens	To Laws
Time	X	1200	1115	1140	1230	1300
Vehicle Moves	X	X	X	X	X	X
Time	X	X	X	X	X	X
Session 2 Time	1300-1530	1300-1530	1300-1530	1300-1530	1300-1530	1330-1530
Final Chopper Transports	To Bishop	X	To Bishop	X	X	To Bishop
Time	1600	X	1530	X	X	1630

TABLE 4

REFERENCES

Crossfield, J.K. 1990, Optimizing GPS Corridor Control Surveys, <u>Technical Papers</u>, Spring 1990 ACSM Conference, Denver, pp. 49-56.

F.G.C.C. 1988, <u>Geometric Geodetic Accuracy Standards and Specifications for Using GPS Relative Positioning Techniques</u>, Version 5.0, National Geodetic Survey.

INCORPORATING KNOWN EXTERNAL DATA IN INFORMATION DIGITIZED FROM CADASTRAL MAPS

Y. Doytsher & B. Shmutter
Technion - Israel Institute of Technology
Haifa , Israel

ABSTRACT

Digitizing cadastral maps creates a "digital copy" of the map, which fits the drawing graphically but contains errors of different sorts, mainly digitizing process errors and inaccuracies inherent in the original drawing. Consequently the derived coordinates of the points defining the parcels are different from the "true" values of the same points which would have been derived by a direct field survey.

The external information which can be added to cadastral maps is of two types; Explicit data (digital values of registered areas of parcels, right of ways etc.); and implicit data (identification of geometric shapes - straight lines, circular arcs etc.).

It is proposed, in order to improve the digital output and to transform it into an accurate "digital map", that all external data be incorporated by means of a semi-automatic identification process of the geometric shapes and their relationships.

INTRODUCTION

Cadastral maps describing land ownership are an important component of any accurate GIS or LIS. As numerical information describing the cadastral maps is rarely available, it must be achieved by digitizing or scanning & vectorizing the graphical maps. The result of such processes are coordinates of the boundary points defining the cadastral parcels which differ from the "true" values as a function of the digitizing process.

The cadastral mapping in Israel is based on the Torrens system (registration of titles). In order to legally validate the digitized information from the cadastral point of view it should be corrected so as to conform to the original information of the cadastral map. The purpose of correcting the coordinates is creating a simulation of cadastral field surveys, in order to eliminate the discrepancies deriving from inaccuracies in the cadastral drawings and errors inherent in the digitizing process itself.

The external information is of two types :-
(1) "Cadastral" Information :
    - Registered area of parcels
    - Right of ways
    - Length of frontages
(2) "Geometric" Information :
    - Closed shapes (parcels)
    - Straight lines
    - Circular arcs
    - Parabolic lines
    - Parallel curves (lines, arcs etc.)

Theoretical aspects of part of these subjects have been discussed by the authors and can be found in [Doytsher 1980], [Shmutter & Doytsher 1987] and [Shmutter & Doytsher 1990].

The main issue in incorporating known external data is forming a relationship between the numerical data resulting from the digitizing process and the different cadastral and geometric conditions. Current solutions based on interactive manual editing, make quite a long and exhaustive process, and are expensive in terms of time and effort.

The proposed solution is based on semi-automatic processes, thus offering the possibility to identify the cadastral and geometric conditions and to connect it to the digital data. Such a relationship enables to solve the resulting condition equations and to achieve the desired purpose, namely - correction of the initial coordinates resulting from a randomly digitizing ("spaghetti") process.

METHOD

The outcome of the spaghetti digitizing process of the cadastral map are arcs, i.e. lines (straight or curved) separating between every two parcels. The extremities of the arcs (startpoint and endpoint) are nodes, - junction points of three or more parcels. In addition, discrete points (centroids), each located within the area of the parcel to which the parcel number is being linked as an alphanumeric attribute, are also picked up.

A list of registered areas of parcels for cadastral maps is available as external data, where the area is given according to the parcel numbers which are unique identification numbers. Numerical values of rights of ways are also linked to the parcel numbers (roads) and so are the lengths of frontages linked to the arcs of the relevant frontages.

The computerized solution deals with separate lines, relationships between lines and with areas. Editing lines is divided in two stages. The first stage is creating continuity between the different arcs and forming continuous lines. The second stage is subdividing the continuous lines into segments according to their geometric shapes.

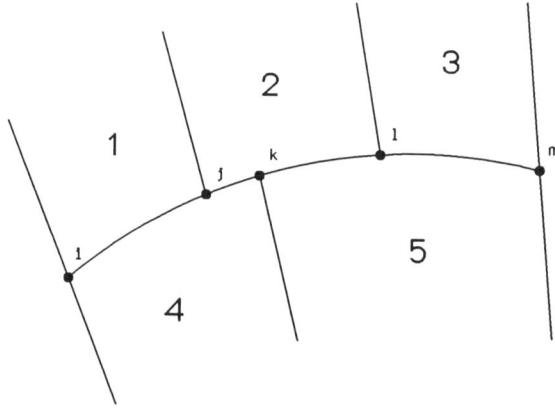

Figure 1. Continuity of arcs

Fig. 1 depicts a part of a cadastral map with the arcs i-j; j-k; k-l; l-m (i-j separating between parcel 1 and parcel 4; j-k between parcels 2 and 4 etc.).

Identification of continuity is carried out in an automatic process, so that in the previous example the result is a continuous line i-j-k-l while the original data includes 4 separate arcs. Identifying arcs as being candidates for creating a continuous line is done according to their locations and directions (differences of coordinates and differences of azimuths smaller than a predefined criteria).

The next step is to subdivide the resulting continuous line of the previous stage into separate segments according to their geometric shapes. The purpose of this procedure is to classify straight lines, circular segments and parabolic curves. In addition, parabolic curves are related to appropriate mathematical polynomial functions according to their curvature.

A straight line is defined as a group of points for which every 3 consecutive points i-1, i, i+1 along the segment fulfills the following condition :-

$$\operatorname{tg} \alpha = \frac{(Y_{i+1}-Y_i)(X_i-X_{i-1}) - (Y_i-Y_{i-1})(X_{i+1}-X_i)}{(Y_{i+1}-Y_i)(Y_i-Y_{i-1}) + (X_{i+1}-X_i)(X_i-X_{i-1})}$$

$$D = \frac{1}{(Y_{i+1}-Y_i)^2+(X_{i+1}-X_i)^2} + \frac{1}{(Y_i-Y_{i-1})^2+(X_i-X_{i-1})^2} \qquad (1)$$

$$\alpha \leq \epsilon \sqrt{D}$$

ϵ being defined according to the accuracy and the scale of the cadastral map.

After identifying the straight lines, the circular segments are to be identified. The latter, if not predefined as such by a suitable code at the input stage, will be identified by their "character". A circular segment is characterized by a large number of consecutive points creating relatively short chords of gradually changing azimuths. The circular segment is therefore defined as a group of points for which every 3 consecutive points i-1, i, i+1 fulfill the conditions :-

$$K_1 \leq \frac{(Y_{i+1}-Y_i)^2 + (X_{i+1}-X_i)^2}{(Y_i-Y_{i-1})^2 + (X_i-X_{i-1})^2} \leq K_2 \qquad (2)$$

$$\alpha_1 \leq \alpha \leq \alpha_2$$

where α_1, α_2, K_1, K_2 are defined as a function of the accuracy and the scale of the cadastral map. The value of α is computed according to (1) and the number of these chords is not less than a predefined number.

All the remaining segments that have not been defined as straight or circular segments are being therefore automatically defined as parabolic curves. Such a definition is not enough and the order of a fitting polynomial line must be determined for each one of the parabolic curves.

It was decided to choose orthogonal polynomials as a suitable approximation for the curves. Using an orthogonal polynomial enables an iterative computing process up to the order of the best-fitted polynomial line. Computing the coefficients of a polynomial in a higher order is based on the coefficients of the previous polynomial (lower order), so that there is no need to start the procedure from the beginning while passing from one order of a solved polynomial to a higher one.

An orthogonal polynomial is defined as :-

$$t = C_0 \Phi_0(s) + C_1 \Phi_1(s) + C_2 \Phi_2(s) + \ldots$$

$$C_i = \frac{\Sigma \, t \Phi_i(s)}{\Sigma \, \Phi_i(s)^2} \tag{3}$$

$$\Phi_k(s) = s^k - A_{k-1} \Phi_{k-1}(s) - A_{k-2} \Phi_{k-2}(s) - \ldots - A_0 \Phi_0(s)$$

$$A_{k-j} = \frac{\Sigma \, s^k \Phi_{k-j}(s)}{\Sigma \, \Phi_{k-j}(s)^2}$$

where s and t are coordinates measured in a local grid system as depicted in Fig. 2 and the value of $\Phi_0(s)$ is defined as 1.

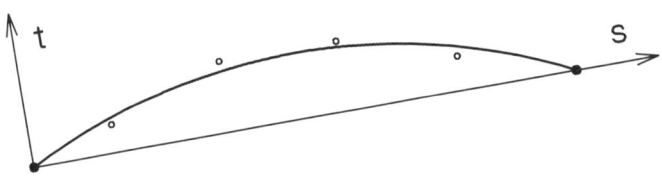

Figure 2. Orthogonal polynomial line

Computing an approximated polynomial line is done in stages. At the first stage a second order polynomial is computed and the differences v_i between the solved line and the original points are evaluated. If both maximal and minimal v differences and their standard deviation are smaller than predefined (as a function of the map) criteria, then that would be selected as the approximated polynomial line. If those conditions are not fulfilled, a third order polynomial line is solved based on the second order polynomial line and so on.

order	v_{min}	v_{max}	$v_{average}$	$v_{m.s.e.}$
1	-1.112	0.767	-0.603	0.622
2	-0.053	0.068	-0.029	0.032
3	-0.050	0.059	-0.028	0.029
4	-0.055	0.026	-0.049	0.053

Table 1. Orthogonal polynomial approximation

Table depicts results of the proposed process. The original line was composed of 12 points picked-up from a map at the scale of 1:2500 (the values in the table are in meters). Based on the numerical values, one concludes that in this case a second order polynomial line is definitively suitable, a third order solution adds only a little and the fourth order will deform the geometric shape.

Every line geometrically determined (in one of the 3 different shapes) defines a separate condition equation. A detailed description of the different condition equations as well as the formulae of coordinates adjustment can be found in [Doytsher 1980] and in [Shmutter & Doytsher 1987]. The purpose of the condition equations regarding the geometric shapes is of a double order: On the one hand smoothing the lines shapes and eliminating discrepancies caused by digitizing process and on the other hand avoiding "distortions" of the geometric shapes which may derive from the adjustment of parcels (from computed to registered area). The process of determining the "geometric" segments is automatic, no manual intervention is needed and the result is condition equations in the number of the defined segments.

Fig. 3 depicts results of this automatic process. The closed figure is the perimeter of a parcel and the black circles are the break-points which have been defined by the proposed method. The closed perimeter was subdivided into 8 separate segments: 4 straight lines (depicted as dash lines), 2 circular segments (wide continuous lines) and 2 parabolic curves (thin continuous lines).

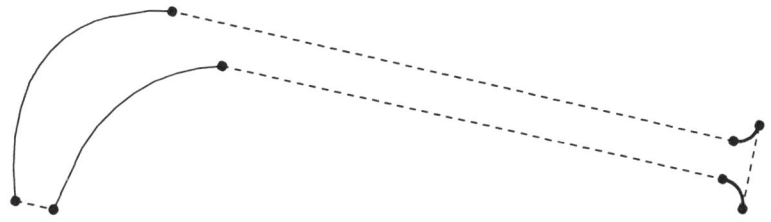

Figure 3. Subdivision of a continuous line

Each parcel within the cadastral map has its registered area A_r. By determining all the arcs defining its perimeter (process which is described in [Shmutter & Doytsher 1990]), the computed area A_c can be obtained from the digitized coordinates. According to cadastral regulations (in Israel), the difference between the computed and the registered areas must not exceed the size of A_p. The difference above the tolerance is :-

$$\delta A = A_r - A_c \quad ; \quad \delta A > A_p \tag{4}$$

The correction to the coordinates of the points defining the parcel will be done according to the formulae presented in [Shmutter & Doytsher 1987].

The process of identifying the parcels and linking the registered areas is automatic, and the result is condition equations in the number of parcels within the cadastral map.

Dealing with lengths of frontages and with parallel lines is principally similar to the process described above. The formulae are detailed in the previous references. Regarding parallel lines, in this stage the process is semi-automatic and manual intervention is needed in order to define pairs of geometric segments in cases of lines which are not straight. At the next stage, an effort to automatize this process will be carried out as well.

PRACTICAL EXAMPLE

As an example to the computation process and to the quality of the results, Fig. 4 depicts a typical cadastral map. The original map is at the scale of 1:2500 and contains 161 parcels, 529 arcs and 2074 points.

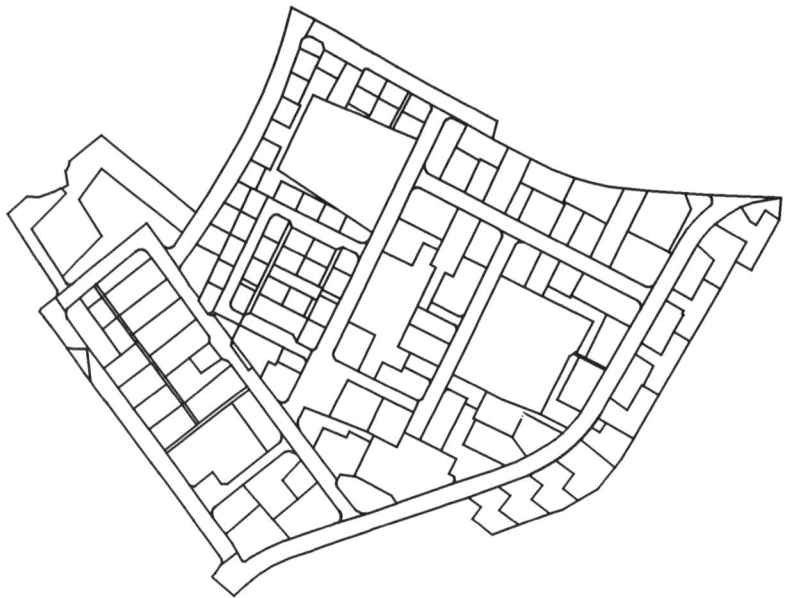

Figure 4. A cadastral map

Considering the number of conditions to be met and the amount of data involved, it is suggested that the adjustment is done in steps and by iterations. Steps mean that each group of conditions is adjusted separately. Iterations mean that the conditions within a group are solved one by one. The coordinates of points resulting from fulfilling one condition equation are subject to the following condition equations. After one pass of solving the different groups of geometrical and cadastral conditions, the process repeats itself until the coordinates of the points remain practically unchanged.

The cadastral block depicted in Fig. 4 was adjusted in two different modes. In the first mode the adjustment was done as a totally separate block, the inner points and points on the perimeter of the block were therefore free to be corrected. In the second mode of adjustment the block was treated as one of a group of cadastral blocks and in this case the perimeter was kept unchanged and only the inner points were corrected.

Table 2 depicts the maximal correction of parcel areas δA (square meters) and the maximal correction of coordinates of points δl (meters), in each one of the iteration stages in the two adjustment modes.

Iteration	Separate block		Group of blocks	
	δA	δl	δA	δl
1	200.66	0.205	200.66	0.408
2	70.75	0.057	92.69	0.212
3	41.17	0.030	65.56	0.152
4	20.60	0.024	55.68	0.130
5	17.84	0.013	51.14	0.093
6	0.00	0.003	35.26	0.072
7			32.89	0.041
8			17.54	0.018
9			20.95	0.016
10			3.07	0.016
11			0.00	0.000

Table 2. Area and coordinate differences

Table 3 depicts the cumulative corrections to the coordinates in the two adjustment modes. The maximal and minimal corrections, the average and standard deviation are given in the Y, X and diagonal (L) directions (numerical values are given in meters).

	Separate block			Group of blocks		
	Y	X	L	Y	X	L
v_{min}	-.371	-.482	.000	-.266	-.490	.000
v_{max}	.244	.429	.499	.764	.500	.787
$v_{average}$	.000	.000	.173	.073	-.048	.211
$v_{m.s.e.}$	.142	.164	.216	.220	.191	.292

Table 3. Coordinate corrections

As foreseen, in the second mode of adjustment, when the block perimeter is not allowed to be corrected, the number of iterations and the numerical corrections of the coordinates are larger than the ones in the first mode. In any case the number of iterations and the size of the corrections are small (standard deviation of about 0.1 millimeter and a maximal correction of about 0.3 millimeter at the original map's scale).

SUMMARY

Adopting techniques of identifying geometrical shapes, together with incorporating external cadastral data, enables to significantly improve the digital data resulting from a "spaghetti" digitizing process of cadastral maps.

The process of adjustment based on external information (geometric and cadastral) transforms the "digital copy" of the cadastral map into a "digital map" of such data. The process is almost absolutely automatic and results in better digitizing output through a relatively simple computing effort.

A practical example depicts the proposed process and its results.

REFERENCES

Doytsher Y., 1980, Numerical Processing of Graphical Cadastral Information, Proceedings of the 16th FIG Congress, paper 304.5.

Doytsher Y. & Shmutter B., 1987, Low Cost Approach to Digital Mapping, Journal of Surveying Engineering, Vol. 113, No. 3.

Shmutter B. & Doytsher Y., 1987, Computerized Solution for Land Subdivision, The Canadian Surveyor, Vol. 41, No. 1.

Shmutter B. & Doytsher Y., 1990, Assembling Closed Polygons, Survey Review, Vol. 30, No. 235.

SOME NOTES ON ADJUSTMENTS, WEIGHTS, AND ACCURACIES

Joseph F. Dracup
National Geodetic Survey, NOS (Retired)
12934 Desert Glen Drive
Sun City West, Arizona 85375

ABSTRACT

Computational practices have not kept pace with technological advances in field surveying despite the availability of computer programs that can handle the least squares adjustments of all but the largest of projects. There are many reasons for this situation and one is certain to be a lack of understanding of what happens in the least squares process and what the results mean. A possible solution to understanding the problem is offered by first giving short descriptions of the general methods of adjustment, i.e., condition equations and observation equations. This is then followed by brief explanations of the sequence of operations the observations and their associated equations go through in an adjustment, supplemented with numerical illustrations to help visualize the procedures. Finally the basic rationale for weights and accuracy estimates is presented.

INTRODUCTION

The last two decades have seen a large improvement in survey accuracies, most of it due to technological advances in instrumentation, yet at the same time, similar advances in computers have not led, in a general sense, to as great an improvement in computations. This is true despite the availability of small computers and programs that can handle the least squares adjustments of all but the largest of projects. Furthermore, there are hand-held calculators with sufficient capacity to do the job for many surveys with only a modest increase in effort than required for these computers. Nonetheless, most surveys continue to be computed in accordance with long used practices. The reasons for the disparagement between the quality of the observations and the quality of the results can be attributed to several factors, with the most important ones being: a reluctance to change and satisfaction with the "rule" methods; little knowledge and consequently some fear of the least squares process, and to some degree, a doesn't matter attitude. Here the intent is to reduce the question of least squares adjustments to a basic level without getting involved in the intricacies of the process. Computers can take care of the intricate as well as the mundane aspects. All that is needed to implement least squares adjustments is a simple understanding of what the method does and how it does it.

METHODS OF ADJUSTMENT

There are two basic methods for making least squares adjustments, condition equations and observation equations. Each of these methods are called by other names which leads to confusion. For example, the method of condition equations is sometimes called adjustment of observations only and adjustment by correlates, while the method of observation equations is often referred to as variation of coordinates, which is acceptable nomenclature, and occasionally, as method of indirect observations and even as condition equations, among others, for both methods.

Condition equations, as the name implies, require setting up equations that must rigorously satisfy the a priori conditions within a network, often called geometric conditions and include those for closures involving triangles, sides, lengths, azimuths, and positions. Observation equations, on the other hand, require equations specifically concerned with the observations themselves, i.e., directions (angles), lengths, and azimuths.

To illustrate each of the adjustment practices and to provide a basis for later discussions on weights and accuracies, several adjustments were made for a short traverse, shown in Figure 1. Selected portions of these computations are compiled as Tables 1.1 through 1.5 and 2.1 through 2.7. Notes concerning their compilation are given below. All tables can be described in simple matrix notation, however the intent here is to provide a fundamental view of what actually happens in the adjustment process to those with no or a minimum understanding of matrices.

Compilation of the tables
General notes concerning the tables follow. Notes pertinent to a specific table are included with the table.

- All tables are for illustration purposes only.

- All numbers shown were truncated from larger ones such that a maximum of four significant figures and a minimum of two decimal places remain, except in some cases where the decimal places are zeros and there only whole numbers are given.

- The lists of observations and preliminary computations from which Tables 1.1 and 2.1 were derived are not given and little information is provided on how any of the tables were computed. Such details are found in (Dracup et al. 1973) for general information and (Hirvonen 1971) for accuracies.

- For Tables 1.3, 2.2, and 2.5 the top line in the horizontal boxes are the Normal Equations (weight matrix for 2.5) and the second line within each box, the Cholesky reduction. The Cholesky method for solving equations is found in (Rainsford 1957).

- Final computations of coordinates are not given for any adjustment.

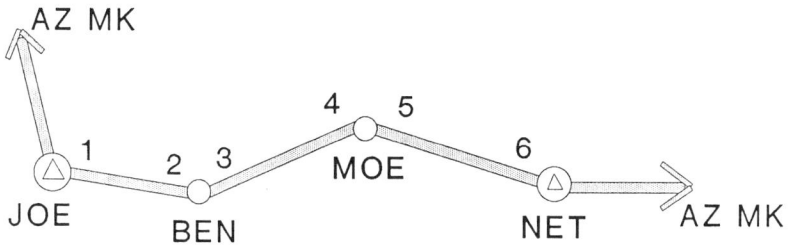

Figure 1:   Sketch of example traverse

Traverse adjustment by method of condition equations
Table 1.1 Condition Equations

1. Azimuth equation $0 = +7.70 + 1.00\ (1) - 1.00\ (2) + 1.00\ (3) - 1.00\ (4) + 1.00\ (5) - 1.00\ (6)$

2. Northing equation $0 = -2.44 - 1.00\ (1) + 0.69\ (2) - 0.69\ (3) + 0.39\ (4) - 0.39\ (5) - 0.05\ (S_1) + 0.10\ (S_2) - 0.13\ (S_3)$

3. Easting equation $0 = -79.71 - 0.93\ (1) + 0.37\ (2) - 0.37\ (3) + 1.39\ (4) - 1.39\ (5) + 3.17\ (S_1) + 2.97\ (S_2) + 3.94(S_3)$

Note: The Northing equation was divided by 10,000 and the Easting equation by 1,000 prior to tabulating above.

Table 1.2 Correlate Equations

	a/p_1	a/p_2	1	2	3	V_1		V_2	
1	9.00	1.00	+1.00	-1.00	-0.93	-1.00		-0.78	
2	9.00	1.00	-1.00	+0.69	+0.37	-4.67		+0.23	
3	9.00	1.00	+1.00	-0.69	-0.37	+4.67		-0.23	
4	9.00	1.00	-1.00	+0.39	+1.39	+10.84		+2.84	
5	9.00	1.00	+1.00	-0.39	-1.39	-10.84		-2.84	
6	9.00	1.00	-1.00			-5.63	Sm	+0.74	Sm
S1	1.00	1.00		-0.05	+3.17	+4.80	+0.07	+6.55	+0.10
S2	1.00	1.00		+0.10	+2.97	+4.39	+0.06	+5.84	+0.08
S3	1.00	1.00		-0.13	+3.94	+6.01	+0.12	+8.27	+0.16

a/P_1 and V_1 are for adjustment #1
a/P_2 and V_2 are for adjustment #2
SE for P_1 = 3.0" for the directions and 1.0" (1:206,265) for the distances
SE for P_2 = 1.0" for the directions and same as above for distances
Sm = correction in unit of distance
a = some convenient number, in this case 1.00.

Table 1.3   Normal equations and forward solution

	1	2	3	$\dot{\eta}$
	+6	-3.18	-4.45	+7.70
	2.44	-1.29	-1.81	+3.14
		+2.32	+2.12	-2.44
		+0.79	-0.29	+2.05
			+39.52	-79.71
			6.01	-12.20

The solution is for adjustment #2 where equal weights are used on the directions and distances. "$\dot{\eta}$" = constant terms of equations.

Table 1.4   Back solution

$C_1 = -0.74$
$C_2 = -1.83$
$C_3 = +2.03$

Table 1.5   Statistics

$\Sigma\dot{\eta}^2 = 163.16$
$\Sigma PV^2 = 163.16$
No. of Condition Eq. = 3
Variance $(SE^2) = PV^2/3 = 54.38$
SE of Unit Weight = 7.37
Σ = sum of

Traverse adjustment by method of observation equations.

For this adjustment the final coordinates for Condition Equations - Adjustment #2 were used in the formation of the observation equations. The weights (P_2) were scaled by the variance of that adjustment (Table 1.5). SE's were 1.0" for the directions and 1:206,265 of the distances expressed in unit of measurement (m) used. By using the final coordinates from the Condition Equations - Adjustment #2, $\Delta N_1, \Delta E_1, \Delta N_2$, and ΔE_2 will be very nearly zero (Table 2.3).

Table 2.1   Observation equations

	P_2	Z_1	ΔN_1	ΔE_1	Z_2	ΔN_2	ΔE_2	N_0	V_2
1	0.01		-62.95	-11.05				-0.78"	-0.78"
2	0.01	+1.00	-62.95	-11.05				+0.00	0.23
3	0.01	+1.00	+62.12	-21.32		-62.12	+21.32	-0.46	-0.23
4	0.01		+62.12	-21.32	+1.00	-62.12	+21.32	+0.00	+2.84
5	0.01				+1.00	+46.46	+16.37	-5.69	-2.84
6	0.01					+46.46	+16.37	+0.74	+0.74
S_1	75.12							+0.10m.	+0.10m.
S_2	79.33		-0.17	+0.98		+0.32	+0.94	+0.08	+0.08
S_3	44.63		-0.32	-0.94		+0.33	-0.94	+0.16	+0.16

Table 2.2   Normal equations and forward solution

Z_1	ΔN_1	ΔE_1	Z_2	ΔN_2	ΔE_2	N
+0.03	-0.01	-0.59		-1.14	+0.39	-0.00
0.19	-0.07	-3.10		-5.95	+2.04	-0.04
	+298.2	-11.55	+1.14	-150.3	+24.35	-3.25
	17.27	-0.68	+0.06	-8.72	+1.41	-0.18
		+16.50	-0.39	+24.35	-87.69	+1.25
		12.44	-0.02	-0.00	-6.45	+0.07
			+0.03	-0.28	+0.69	-0.10
			0.17	+1.62	+2.35	-0.50
				+234.6	-10.36	+1.08
				10.96	+0.94	-0.00
					+137.2	-2.06
					9.10	-0.00

Table 2.3 Back solution

$Z_1 = +0.23$

$\Delta N_1 = -0.00$

$\Delta E_1 = +0.00$

$Z_2 = +2.84$

$\Delta N_2 = +0.00$

$\Delta E_2 = +0.00$

Table 2.4 Statistics

$\Sigma \ PN^2_\circ = 3.30$

$\Sigma \ N^2 = 0.30$

$PN^2 - (N \times N) = 3.00$

$\Sigma \ PV^2 = 3.00$

No. of Condition Eq = 3

Variance $(SE^2) = PV^2/3 = 1.00$

SE of Unit Weight = 1.00

Table 2.5 Weight matrix for accuracies

	N_{1x}	E_{1x}	N_{2x}	E_{2x}	L1-2	Az1-2
ΔN_1	+1 +0.05				-0.32 -0.01	+62.12 +3.59
ΔE_1	+0.00	+1 +0.08			-0.94 -0.07	-21.32 -1.51
Z_2	-0.02	+0.01			-0.00	-1.57
ΔN_2	+0.04	-0.00	+1 +0.91		+0.32 +0.01	-62.12 -2.56
ΔE_2	-0.00	+0.05	-0.00	+1 +0.10	+0.94 +0.05	+21.32 +1.37
$\sqrt{\Sigma x^2}$	0.07	0.09	0.09	0.10	0.09	5.12

Table 2.6 Accuracies-Standard errors

$N_1 = \pm\ 0.07m$

$E_1 = \pm\ 0.09m$

$N_2 = \pm\ 0.09m$

$E_2 = \pm\ 0.10m$

$L1-2 = \pm\ 0.09m$

$AZ1-2 = \pm\ 5.12"$

Table 2.7 Error ellipse data

Sta BEN $Q_{xx} = 0.00$ SE - Major Axis = 0.09 $Q_{xx} = \Sigma N_x^2$
        $Q_{yy} = 0.00$ SE - Minor Axis = 0.07 $Q_{yy} = \Sigma E_x^2$
        $Q_{xy} = -0.00$ $\theta = -82.3°$ $Q_{xy} = \Sigma(N_x E_x)$

Sta MOE $Q_{xx} = 0.00$ SE - Major Axis = 0.11 The above
        $Q_{yy} = 0.01$ SE - Minor Axis = 0.09 quadratic
        $Q_{xy} = -0.00$ $\theta = -75.2°$ equation
                                                 is solved
                                                 for the
                                                 roots λ_1, λ_2.
                        Major axis = $SE\sqrt{\lambda_1}$.
                        Minor axis = $SE\sqrt{\lambda_2}$.
θ = major axis rotation angle; SE = SE of unit weight

<u>Synopsis of an adjustment</u>. In a general way adjustments of condition equations involve steps 1-6, 10, and, for observation equations, steps 1-10, regardless of whether computers are used or not.

1. Gather the observational data, determine the weights $P = 1/SE^2$ and make necessary preliminary computations.

2. Prepare equations and tabulate in prescribed format for computation of normal equations and "Vs". See Tables 1.1, 1.2, and 2.1.

3. Compute normal equations and forward solution. See Tables 1.3 and 2.2.

4. Compute back solution. See Tables 1.4 and 2.3.

5. Substitute results of back solution in Tables 1.2 and 2.1 and compute the residuals or "Vs" = corrections to observations.

6. Compute statistics (Tables 1.5 and 2.4) from Tables 1.3, 1.2, 2.2, and 2.1.

7. For observation equations accuracies are determined by extending Table 2.2 to include the weight or border matrix and compute the weight coefficients $(\Sigma x^2)^{1/2}$. See Table 2.5.

8. Compute position accuracies = SE $(\Sigma x^2)^{1/2}$. See Table 2.6. SE is from Table 2.4.

9. Compute error ellipse data from Table 2.5. See Table 2.7.

10. Correct observations by the "Vs". For condition equations use the adjusted observations to make the final computations and determine the adjusted coordinates. For observation equations, correct the preliminary coordinates for the shifts ΔN_1, ΔE_1, ΔN_2, and ΔE_2 from the back solution (Table 2.3) to obtain the final coordinates and using these values, compute the adjusted azimuths and distances and compare with the corrected observations.

Advantages and disadvantages of each method and other considerations

Reference surfaces. Adjustments can be made on a local plane, most conformal projection grids such as the state and UTM systems and on the ellipsoid. The first computations produce plane coordinates and the last, latitudes and longitudes.

Formation and solution of equations. Forming condition equations for triangulation and trilateration can be complex and since these methods are rarely used today, no further reference is made to them in this paper. Condition equations for traverses, however, are fairly simple to set up, consisting only of equations to distribute the azimuth and position closures. See Table 1.1 for a partial example. Observation equations, on the other hand, require equations for each observation included, i.e., for directions or angles, lengths, and azimuths. See Table 2.1 for a partial example.

Without getting into specifics, the number of condition equations for a traverse such as that shown by Figure 1 is three, regardless of the number of new stations involved. The number of condition equations is also the number of normal equations that need to be solved, but this is not the case for the observation equation method. There, the number of normal equations is equal to twice the number of new points, as a minimum. See Tables 1.3 and 2.2. The solution shown by Table 2.2 is the case where directions rather than angles were observed and one additional normal equation per new point is required. The number of condition equations is often called Degrees of Freedom or Redundancies and also comes into play in computing the variance for both methods (Tables 1.5 and 2.4). In summary, condition equations almost always require the set up and solution of fewer equations (often much fewer) than for observation equations, yet the latter method is preferred where computers are available, because of generally easier programming.

Residuals, final computations, and checks on adjustments. Residuals or "Vs", the corrections to the observations, are the most important end product of the least square process (Tables 1.2 and 2.1). As cases in point, the final coordinates for condition equations are computed from the adjusted observations. For observation equations, the acceptability of the adjusted coordinates are totally dependent on the agreement of inversed azimuths and lengths with the adjusted observations. Such coordinates should never be accepted at face value as some users are prone to do, because of the possibility that erroneous data got into the process. Also, ΣPV^2 (Sum of the weights (P) times V^2) is used in all statistic examinations (Tables 1.5 and 2.4), as well as all accuracy evaluations (Tables 2.6 and 2.7)

Condition equations are easy to set up when all equations, weights, and residuals are expressed in the same unit, usually seconds. This is due to the fact that some equations involve corrections to both angles and distances (Table 1.1). Thus corrections to lengths expressed in angular measure must be converted to the length unit by the formula $ds = Sm = Sds"/206,265$ (Table 1.2). Such is not the case for observation equations since each type of observation has a specific equation (Table 2.1). However, observation equations for length can be so treated as to produce residuals in seconds, if desired.

Evaluation and analysis of data. Condition equations provide the best pre-adjustment evaluation of the observations, in as much as determining the azimuth and position closures is required to set up the equations. There is little else that can be done to evaluate traverses, at this point. Similar computations are suggested for observation equations as well, primarily as an aid to locating mistakes and blunders in the data, or if for nothing else to obtain good preliminary coordinates. However, some users prefer to let the adjustment process locate problems and for the little time it takes to make most solutions, it's difficult to argue against the practice. Nevertheless, one must note that the fundamental criteria for traverse accuracy standards are the closures in azimuth and position.

During the course of solving the condition equations, a continuous view of the buildup in PV^2 is available, if made part of the computer output. Specifically, the sum of the squares of the "$\hat{\eta}$" column Cholesky terms equals ΣPV^2 (Tables 1.3 and 1.5). Note for this example that over 90 percent of ΣPV^2 is attributed to the closure for the Easting equation. Similar information is not easily derived in observation equation adjustments.

For post adjustment analyses, condition equations offer little more than an examination of the residuals and a few statistical tests common for both methods that are often inconclusive. The best analyses of the adjusted results are accuracy estimates in the form of standard errors for the coordinates, lines, distances, azimuths, and error ellipses available from observation equation solutions. The availability of these estimates is a major factor in favoring the use of observation equations over conditions.

Computer programs. Few programs are available for adjusting traverse by the condition equation method, yet it is a viable approach and in some respects a better method than observation equations. Fewer equations need to be solved, iterations of the equations and solutions are rarely if ever necessary, final coordinates are computed from the adjusted observations, and the subsequent assurance of their reliability are among the more obvious reasons. One program was developed by (Holdahl and Dubester 1971) for use with the state plane coordinate systems (SPCS) that included the necessary terms to provide adjusted results closely approximate to those obtained from a computation on the ellipsoid. This program can be used for most NAD 83 SPCS, with little modifications. The major problem in setting up an all inclusive program for condition equations is the need to develop a subroutine for selecting the traverse links over which the equations are formed for multi-line and loop type networks. Previously this was done by visual inspection and included in the input data (Dracup et al. 1973).

At the present time, least squares adjustment programs generally available to the public are usually referenced to a local plane, use the observation equation method, and feature error ellipse information (Curry et al. 1989). Most use angles rather than directions, which limits the number of equations in a solution to two for each new point, rather than the three per point required when directions are used. The results (angles versus directions) are not identical, but the differences are rarely of any consequence and it is seldom worth the additional effort to bring angle adjustments into exact agreement. SPCS can be adjusted on a local plane provided the angular, scale, and elevation reductions (second term, scale factors, and ellipsoid) are taken care of in some fashion. Solutions are iterated until some prescribed tolerance in the coordinate shifts or ΣPV^2 is reached. Either approach is satisfactory, albeit the latter may require more iterations. Many programs are three dimensional (3D) in a sense, because vertical angles for elevation determinations and slope reductions can be included as part of the adjustment process. Such programs are useful, more for a matter of convenience than anything else, since very little or no improvement in the coordinates or elevations result from the simultaneous computation.

WEIGHTS

Weights in the simplest form can be described as the means for including observations of unequal worth in a computation. Properly employed the use of such equalizers provides good assurance that the final results represent the best that can be obtained from all the measurements. As a general rule, the weight "P" = $1/SE^2$, where SE is the standard error of the observations, based on the observations themselves or as is more likely the case, predetermined standard errors derived from large series of observations made following specifications that approximate those used in the particular survey. No great precision in the SEs are required. It is immaterial whether a SE is 3.0" or 3.2".

Relation to variance. In theory, a properly weighed solution will give a variance of 1.00 or ΣPV^2 = number of condition equations (Tables 1.5 and 2.4). This rarely happens in practice and especially in those cases where the observations are constrained to the fixed control, such as the example. The variance can always be made to equal 1.00 by running a second solution where the weights are divided by the variance of the first computation. Nothing else changes, the residuals (Vs) remain the same, as do the accuracies. To demonstrate this point in the observation equation adjustment (Tables 2.1-2.7), the original weights were divided by the variance from the condition equations (Table 1.5). The square root of the variance is known as the Standard Error of Unit Weight.

Reviews of weighted adjustments. Results of least squares adjustments should always be reviewed to assure that they fall within those expected for the survey specifications used. To illustrate this point, occasions arise where the observations and their assigned weights conflict with the constraints imposed by the fixed control. The following describes such an example. For condition equation adjustment #1 (Tables 1.1 and 1.2 only), the weights were based on SEs = 3" (1:68,755) for the directions and 1" (1:206,265) for the distances and the residuals are shown by Table 1.2. The large corrections on directions were caused by the weight differential of 9:1 favoring the distances in an east-west survey, where the large easting closure (Table 1.1) requires significant corrections to the measured distances regardless of the worth of the directions. Equal weights were used for adjustment #2 and a much smoother group of residuals resulted (Table 1.2). Now the question arises, which adjustment is better?. Some will argue with considerable support for their views, that if the weights are representative of the accuracies of the observations, they should not be changed to satisfy extraneous conditions such as closures and hence the first adjustment is best. Others would chose the second for that very reason, the results are in line with the closures (Table 1.1). And this choice is borne out to some degree, by the accuracy estimates discussed later. Which one is best is secondary to the issue raised here that a careful review of the results should always be made and never blindly accept the data because it came from a least squares adjustment.

Used as constraints. Weights can be used to partially or fully constrain (hold fixed) directions, angles, lengths, and azimuths for both adjustment methods by assigning appropriate values to the observations that will satisfy the particular requirement and for observation equations the same can be done for the coordinates, as well. Care must be exercised in such specialty weighting situations to assure that problems do not arise elsewhere as a result.

ACCURACY ESTIMATES

Accuracy estimates came into more general use with the advent of computers. Prior to that time observation equation adjustments which could provide the information were considered too laborious and costly, except for a few special projects, and adjustments by condition equations were the rule. Today there is considerable interest in such estimates and that is a good trend. But, there is also a need for a better understanding of them and that is a crux of a problem.

The what and how of accuracies. Accuracy estimates are a reflection of the adjusted results as derived from a mixture of observations, weights, survey geometry (network design for traverse), and fixed control constraints that comprise a least squares computation and not, as often thought, the observations alone. When all elements are favorable, the estimates are reliable and meaningful. However, when one or more of the elements, other than the observations is bad, for whatever reason, the estimates can deteriorate rapidly. They do nevertheless represent the results as adjusted. In all instances, the accuracies are relative to the fixed control, except for line accuracies.

The following accuracy estimates can be obtained from a weight or border matrix illustrated by Table 2.5. Circular errors can be computed from the same matrix, as well, but are rarely of interest today.

- Coordinates - expressed as SE of each coordinate (N,E) (Table 2.6)

- Line - expressed as SE in length and azimuth (Table 2.6). Also, can be given in form of SE of relative coordinates (N,E) (no table)

- Error ellipses - derived as SE ellipse (39.35 percent confidence region). Usually shown at the 95 percent confidence region. This is done by multiplying SE axes (Table 2.7) by 2.45. Also, can be shown as a line ellipse (no table).

Error ellipse computations are included in many computer programs for obvious reasons, they give a picture of the accuracy estimates. Line accuracies are generally considered better by many geodesists because they give information between specific points.

Accuracy evaluations. Condition equation adjustments #1 and #2 were discussed previously in regard to the weights used, here the examination is continued further. Adjustment #2 accuracies are given by Table 2.6 and error ellipse data by Table 2.7. The constructed error ellipse is elongated in the east-west direction, with a ratio of the axes of 0.8. For adjustment #1 the corresponding information is $N_1 = \pm 0.19$, $E_1 = \pm 0.09$, $N_2 = \pm 0.20$, $E_2 = \pm 0.10$, $L = \pm 0.08$, and $A\dot{z} = \pm 10.7"$. For the ellipses: sta BEN axes = 0.19, 0.09, and O= -10 ; sta MOE axes = 0.21, 0.10, and O = -10 . The constructed error ellipse is enlongated in the north-south direction with a ratio of the axes of about 0.5. The two sets of accuracies reflected the adjusted results, i.e., for adjustment #1, northings distorted by excessive bowing of the traverse due to the large easting closure constraint on the low weighted angle observations and for adjustment #2, the large easting closure is primarily distributed on the lengths.

One final note: with the ease that adjustments can be made on computers, there are no reasons why additional computations should not be made whenever questions arise about the adjusted data. It is far better to do a little more work at the beginning rather than try to patch up problems later.

REFERENCES

Curry, S., R. Sawyer, and J.M. Anderson 1989, Star*Net: Rigorous 2D and 3D Survey Network AdjustmentProgram, <u>Surveying and Cartography</u>, Vol. 5, pp. 269-277, 1989 ASPRS/ACSM Annual Convention, Baltimore, Maryland.

Dracup, J.F., C.F. Kelley, G.B. Lesley, and R.W. Tomlinson 1973, Surveying Instrumentation and Coordinate Computation Workshop Lecture Notes, ACSM, pp. 93-200.

Hirvonen, R.A. 1971, <u>Adjustments by Least Squares in Geodesy and Photogrammetry</u>, pp. 165-169, Frederick Ungar Publishing Co., New York.

Holdahl, J.H. and D.C. Dubestor 1971, A Computer Program for Traverse Adjustment Using Plane Coordinates, ACSM Proceedings, ASP/ACSM National Convention, Washington, D.C.

Rainsford, H.F. 1957, <u>Survey Adjustments and Least Squares</u>, pp.74-96, Constable and Co., Ltd., London.

TAX MAPPING ON THE CONNECTICUT COORDINATE SYSTEM WITH SCALE IMPROVEMENTS, USING APPROXIMATELY SCALED AIR PHOTOGRAPHS, SURVEYORS' MAPS, AND QUADRANGLE SHEETS

Spring 1991 Baltimore Convention Paper No. 202

C. Roger Ferguson
Civil Engineering Department
University of Connecticut
Storrs, CT 06269-3037

Phillip G. Caron
Towne Engineering, Inc.
P.O. Box 162
So. Windham, CT 06266

BIOGRAPHICAL SKETCHES

Roger Ferguson has been associated with the surveying profession since 1956, when he found a summer position as a chainman-rodman while working toward a college degree in accounting. He has earned bachelors degrees in Accounting and Civil Engineering and a masters degree in Environmental Engineering. From 1972 to 1989 he has pursued the dual careers of surveying and engineering teacher and surveying and engineering consultant. Commencing in September, 1989, he has been a full time lecturer at the University of Connecticut. He has been active in professional societies as the chairman of both the Continuing Education and the Scholarship Committees for the Connecticut Association of Land Surveyors (CALS), a director of the New England Section ACSM, and the Connecticut representative to the NSPS Board of Governors.

Phillip Caron, L.S.I.T., is manager of Technical Services at Towne Engineering, Inc., South Windham, CT, where he is responsible for computer hardware and software technical support and scheduling of field crew activities. Active in the surveying profession since 1979, he is one of the three members of the CALS Continuing Education Committee which has created the Surveyors Training Seminar Series. The Committee also simultaneously administers a professional level seminar program. He is also an evening instructor of surveying at Thames Valley State Technical College, Norwich, CT. His specialities include land records research, boundary retracement techniques, surveying computations, and computer aided design and drafting.

ABSTRACT

The Town of Columbia, CT has experienced substantial development in the past few years and has decided that their 400 scale tax maps are no longer usable. The tax assessor has decided to have a new set of maps prepared at 200 scale, except for the town center and lake areas, which will be drawn to 100 scale. The present tax maps are on approximately scaled photographs and the new photographs purchased by the tax assessor are also approximately scaled, i.e. they are not orthophotos.

In considering this project, the project team decided to attempt to provide better scaled tax map mylars by combining available digitized U.S. Geological Survey quadrangle sheets on the Connecticut Coordinate System with surveyors' filed maps and the approximately scaled photographs. The primary tool utilized for this project is an IBM compatible 386 computer with AutoCad and DCA software.

Some of the surveyors' maps are on the Connecticut Coordinate System, but most are not. The AutoCad and DCA software will be utilized to do the map rotating and enlarging-reducing which would have been very labor and time intensive using hand drafting methods.

Utilization of the Connecticut Coordinate System will also prepare the Assessor's Tax Maps for entry into a GIS/LIS system when the Town decides to implement a GIS/LIS.

This paper will describe the process utilized, talk about the problems encountered, and indicate how successful we have been in producing better scaled maps than are generally produced from the approximately scaled air photographs.

INTRODUCTION

The Town of Columbia, CT currently has a population of approximately 4000 and an area of 21.8 square miles. Its primary industry is agriculture, although it does have a few commercial and industrial businesses. There are two lakes in town, one of which has many summer cottages on its shores. Many of its residents commute to urban areas to work.

The first tax maps prepared for the Town of Columbia were completed in the mid 1960's by a large, out of state mapping firm. Since there were many acres covered with large tracts of land, most of the maps were drawn to a scale of one inch equals 400 feet (1" = 400' or 400 scale). In one commercial area, the maps were 200 scale, and in the village center and around the largest lake, 100 scale maps were drawn.

For the first few years after the tax maps were prepared the out of state mapping firm provided annual updating of tax maps. The annual updating reflected new lots created, changes in lot lines due to property exchanges, and changes in lot lines due to new information provided on surveyors' maps completed during the current year. This proved to be unsatisfactory economically due to the travel distance required for the mapper and in terms of scheduling the updating work and answering the mapping questions that surfaced between updates. In 1973, the Columbia assessors decided to engage a local land surveyor as a mapping consultant who would be available if needed during the year and who would do the annual tax map updating. One of this paper's coauthors, Roger Ferguson, was chosen as the mapping consultant and has served the Columbia assessors in that capacity since then.

During the 1980's many large tracts of land were subdivided, creating a very large number of new lots and a problem for the tax assessor and the tax mapping consultant. Simply stated, the minimum lot size of 40,000 square feet is small enough that the required identifying information about each lot will often not fit within the available space on 400 scale maps. After consultation with the mapping consultant, the assessor decided that new, larger scale maps were necessary. The Board of Selectman and Board of Finance agreed and funds were allocated to prepare new tax maps over a three year period. Because of his long association with the Columbia tax maps and consequent familiarity with the town, Ferguson was chosen to prepare the new tax maps.

HOW SHOULD THE MAPS BE PREPARED?

The existing mapping was done on approximately scaled air photos. The tax maps themselves are mylar tracings which show lot lines, highway lines, street names, lot dimensions (lengths only), lot areas, parcel identifying numbers, and subdivider's lot numbers, but no buildings or other planimetric information. Since the air photos used were not orthophotos, there are numerous areas where distances scaled on the maps are significantly incorrect.

Although it was known that orthophotos would be more expensive, it was decided that it was worth investigating to see if it would be worth paying the price for maps that would not have so many areas where even approximate scaling could not be confidently done. The result of the investigation was that the orthophoto cost was deemed to be too great for a small town when weighed against the benefit derived from being able to scale confidently at 200 scale.

The Connecticut Department of Environmental Protection (ConnDEP) has air photos taken of the entire state every five years. In exchange for a favorable fee for that service, the successful bidder for the air photo contract is also awarded the franchise to sell the air photos to the public. Consequently, approximately scaled air photos are available for a very reasonable cost to many potential users. The initial conception of how to create the new 200 scale maps was to purchase the 1985 ConnDEP flight photos and recreate the tax maps in much the same manner as they were previously done, but now basically at 200 scale, rather than 400 scale. The air photos would provide the planimetric data, such as stone wall boundary locations, water course locations, locations of property boundaries such as cleared fields vs. wooded land, highway locations, and building locations which is useful in orienting existing surveyors' maps to their proper location within the town. The old 400 scale maps would be useful too, but the hope was that some of the scaling problems could be eliminated. To accomplish this, they could not simply be photoenlarged.

About three quarters of the approximately 2000 parcels of land in Columbia are shown on about 600 usable surveyor's maps filed in the town clerk's office. The prospect of reducing the 1500 mapped parcels, by hand graphic methods, from the filed map size to 200 scale was not an appealing one to the mapping consultant, most particularly in light of his knowledge of some of the capabilities of computer aided drafting (CAD), and particularly AutoCAD.

The first thoughts about utilizing AutoCAD were to simply enter each filed survey map into AutoCAD through a COGO package, then utilize AutoCAD to plot each map to the required scale (200 or 100). Each plotted map would then be "best fit" to the air photo and its attendant scale distortions. Further thoughts about the process led to the idea of improving the tax maps considerably by using the 200 scale air photos as a guide, but consolidating the filed survey maps into an AutoCAD "mosaic' which would have the true scale survey map dimensions as a base from which unsurveyed parcel dimensions could be scaled. Additional conjecture led to the thought that by utilizing AutoCAD digitizing techniques the air photos might be brought closer to scale by comparing the relationship between various identifiable objects on them to the U.S. Geological Survey 1:24,000 quadrangle sheets. Then, why not utilize the Connecticut Grid System (CGS) tick marks on the quadrangle sheets to control the placement of all of the parcels in Columbia? This was a particularly attractive idea in that coauthor, Ferguson, had surveyed two large contiguous blocks containing many parcels of land using the CGS as the coordinate system for the several surveys. One of the blocks was completed over a several year period, contained fourteen (14) interlocking control survey traverse loops and fifty (50) parcels of land, and covered about 900 acres. The second block contained 30 parcels of land and covered about 180 acres. The two blocks of surveys contained frontage on the three major state highways traversing Columbia, so the state highway maps could be easily rotated on to the CGS. With the state highways on the CGS many of the other filed survey maps could be as easily rotated on to the CGS from the magnetic north, assumed coordinate systems they were prepared on. Add to these the few surveys prepared by Ferguson and other surveyors on the CGS in other areas of town and there is a small, excellent framework upon which to begin attaching other survey maps, and eventually unsurveyed parcels.

In that Ferguson's exposure to and knowledge of AutoCAD was as a surveying-engineering manager, with no hands on experience, it was necessary to enlist the aid of a surveying-computer expert. Coauthor Phillip Caron is extremely well versed in computer hardware and surveying and engineering software and is Manager of Technical Services at Towne Engineering, Inc. He agreed to lend his computer expertise to the project and Towne Engineering, Inc. agreed to rent computer hardware and software to Ferguson. The project could now begin.

Consultations on the direction for the mapping project resulted in ideas for some different approaches than initially envisioned. Utilization of the printed quadrangle sheets was discarded in favor of purchasing digitized quadrangle (Columbia and Willimantic, CT) disks from Graphic Information Systems of Houston, TX. The disks purchased contained all quadrangle sheet information except for contours and building locations and were purchased with the CGS, rather than latitude and longitude as the coordinate basis. These two quadrangle sheets were merged to form the base drawing for all of the Columbia tax maps. Many maps can be placed into the base system by simply matching planimetric data from the quadrangle disks to the same data (streams, road frontage geometry) on the survey maps.

When completed, the tax maps can be plotted on mylar to provide hard copies for assessor's and general public use, but can also be stored on floppy disks and/or 1/4" high speed data cartridges. Future maintenance and updating of these files will be done in the AutoCAD vehicle. If new mylar plots are not desired each time a change is made, paper plottings can be made and the changes traced on to the existing assessor's mylar master copy. This will be accomplished much more easily than the current method of trying to fit a well scaled surveyor's map to the poorly scaled mylars generated from poorly scaled air photos. The changes will also be stored in the master floppy disk and data cartridge files.

The AutoCAD produced and stored product, referenced to the CGS, will also provide the basis for a comprehensive Land Information System (LIS) or Geographic Information System (GIS). AutoCAD is widely use in mapping operations, and is in the opinion of many, the industry standard CAD package. Many available GIS/LIS packages utilize AutoCAD for their graphic capabilities and those that do not should be able to accept AutoCAD generated maps through a transfer (DXF) file. Once completed then, these tax maps should be capable of providing the coordinate and parcel framework for any GIS/LIS.

The opinion of the coauthors at the time the plans for the project were completed was that utilizing the AutoCAD system would allow production of a far better product, faster, and with great potential for expanded use, at a comparable price to the product that could be produced utilizing traditional manual drafting techniques.

PREPARING THE TAX MAPS

The basic scale for the new tax maps is one inch equals two hundred feet (200 scale). A few maps around the largest lake and in the town center will be prepared at 100 scale. The town was broken into a series of eighteen inch by twenty-seven inch (18" x 27") panels which define the limits of coverage of each sheet. See Figure 1. The maps will be produced on standard 24" x 36" sheets. This will allow room for some overlapping information to be plotted on any sheet where it will make the information contained on that sheet clearer. For example, if the sheet match, sheet coverage line cuts through the middle of the frontage of a 150 foot frontage, 40,000 square foot lot, the entire lot could be shown on each of the sheets. The parcel identifier and dimensions can appear on one of the sheets, and a reference to that sheet and parcel identifying number can appear on the other sheet. Forty-four sheets will be required to cover the town. Each of the sheets is drawn with north straight up, as most users are accustomed to viewing maps in this manner. Note that it was quite easy to set up and shift panel arrangements in AutoCAD to determine the one which utilized the least number of sheets and had the least number of sheets having very small areas of the town covered on them. At this writing, the number of and arrangement of the 100 scale maps has not yet been decided upon. There will also be an index map of the 100 scale maps. The parcels shown on the 100 scale maps will contain their dimensions and parcel identifiers thereon. The 100 scale map parcels will also be drawn on the 200 scale maps, but without dimensions, and with reference to 100 scale map parcel identifiers and map numbers.

Each of the existing filed maps to be utilized is created in digital format by entering bearing and distance, or angle and distance information, utilizing the DCA civil engineering software package within AutoCAD. Maps referred to the CGS are entered on that coordinate system, others are entered on a randomly assumed coordinate system. Each map is stored preliminarily in a separate drawing (DWG) file corresponding to its Columbia Land Records drawer and map file number. This allows easy inventory of the maps that have been created for later insertion into their appropriate final sheets. Those maps that are on the CGS are stored with the letter "c" preceding their drawer and map number. The two large blocks of contiguous lots on the CGS were created as large drawing files labeled with the file number of one of the large parcels contained within the consolidated drawing. The largest block of lots stretches over six (6) maps.

Each of the maps on the CGS are inserted into the final drawing sheets and noted with a highlighter on the working copy of the existing tax maps.

Next the Connecticut Department of Transportation (ConnDOT) right of way maps are entered as drawings on the CGS where possible, on assumed coordinate and bearing systems where not. The assumed system highway maps are plotted to 200 scale and overlaid on the plotted digital quadrangle sheets to obtain the best graphic fit, then entered into the final drawing files by digitizing coordinates of critical points on the quadrangle sheet base drawing and rotating the highway map file to conform to those coordinates.

Then each of the individually stored, filed maps on assumed coordinate systems is consolidated with adjacent filed maps by computer rotation techniques. This creates a "mosaic" of filed maps previously on unrelated random systems which can be matched to CGS system maps or state highway maps where they abut. Where these "mosaics" do not abut state highways or CGS maps they are plotted at 200 scale and physically matched to 200 scale base map plottings to obtain the best fit to roadway geometry and planimetric features, such as streams or utility easements, which show on the quadrangle sheets. Where the "mosaics" cannot be matched to planimetric features on the quadrangle base maps, they are matched to such features on the approximately 200 scale air photos by physically overlaying them. The air photos are then digitized in the vicinity of the subject parcels to conform to planimetric information which shows on both the air photos and the quadrangle base maps. Finally, the "mosaics" are inserted into the individual new tax map sheets through essentially a double translation-rotation.

Individual filed maps which are not contiguous with other filed maps are handled in much the same manner as described above, singly, rather than in "mosaic" groups.

The final group of lots to consider are those which have not been surveyed and recorded in the land records. For these lots, the existing tax map configurations will be used, but they will be fit between the previously entered survey mapped parcels by digitizing the old tax map and forcing the parcels to enlarge or shrink as a group within the space remaining for them on the individual new tax map sheets.

Lineal dimensions will be shown for each parcel. Parcel area and cross references to filed map numbers in the land records will also appear on the final maps. Subdivision lot numbers will be noted on each subdivision parcel.

When it appears that each lot has been accounted for, a set of check plottings will be made. On these plottings the new parcel numbers will be assigned. At this stage, all areas, lengths, subdivision lot numbers, and land records file number references will be checked and highlighted on the check plottings.

With the addition of the parcel identifiers and any corrections necessary the tax maps will be nearly ready to plot. A standard set of legends, assessor's notes and information, the Town Seal, border, north arrow, and title was created to fit around the parcels contained on each map. Shifting position of various elements of each drawing to give it balance, centering, and general eye appeal is easily accomplished in AutoCAD.

Maintenance and use of the final drawing files was carefully considered. Various elements of the drawings are stored on individual layers so that plottings can be made for various purposes. Plottings can be made showing parcel lines only, showing parcel lines and easement lines, showing parcel lines and parcel identifiers, showing all available information, or showing any combination of elements desired.

FUTURE MAINTENANCE AND UPDATING

As previously discussed, future updating and maintenance will be in the AutoCAD vehicle. Changes will be stored on the floppy disks and the data cartridges. The assessor will have the flexibility of deciding whether new mylars should be produced when changes are made, or whether interim changes should be hand traced on the mylars from a paper computer plotting. Since Columbia is a small town,

it may be that new mylars will be desired only at
predetermined intervals such as annually or every two
years. The option of requiring subdividers to submit line
work on disks in DXF format for insertion into the tax maps
can also be considered. This will minimize the amount of
work required of the Town Mapping Consultant in updating
the tax maps.

HARDWARE REQUIREMENTS

The new Columbia tax maps are being produced on NEC
Powermate 386 machines running MS-DOS Version 3.30 at 16
Mhz. Each workstation is equipped with AutoCAD Release
10.2 enhances with DCA Civil Engineering Software Version
10.0.

Digitizing was accomplished with CALCOMP Series 9100
digitizers.

Data file storage is on 1/4" high speed data cartridges and
on 5 1/4" floppy disks.

Final hard copy mylars are produced on a CALCOMP 1043GT
Plotter, using black ink on 3-mil mylar.

CONCLUSION

In that this is an ongoing project there is the possibility
of a follow up paper discussing more findings, citing final
numbers of worker and computer hours required to complete
the project, and documenting the scaling improvements
achieved.

Some conclusions are warranted from the work completed to
date. The process being utilized is producing a much
better product than the initially proposed 200 scale air
photo based maps.

The distortions that would have been created by forcing the
survey maps to conform to the approximately scaled air
photos have been eliminated. In that there are more
surveyed parcels than unsurveyed parcels, the elimination
of this distortion provides a better basis for fitting the
unsurveyed parcels in the remaining area than the air
photos. The scaling distortions in the unsurveyed parcels
are thus also minimized.

Updating of the tax maps will be much easier from now on,
since there will be no more traditional graphic reduction
of maps and subsequent fitting of them into improperly
scaled areas.

Although a GIS/LIS is not envisioned in the near future for
Columbia or many other small towns, the CGS system
framework is in place and the locations of the boundaries
of all properties in the town are referenced to that
framework.

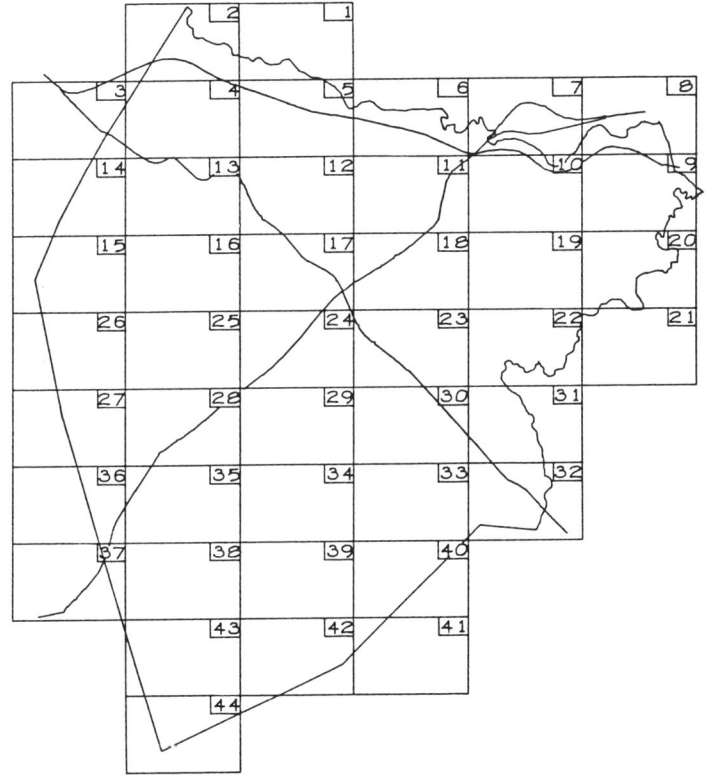

FIGURE 1.
200-SCALE SHEET INDEX
SHOWING STATE HIGHWAYS

W A S H M O R E
WASHINGTON MONUMENT RESURVEY EXPEDITION

George E. Gary, L.S.
Gary Land Surveying
40,001 Carmel Valley Road, Greenfield, CA 93927

BIOGRAPHICAL SKETCH

George E. Gary retired from California Department of Transportation in 1972 and at the same time moved from Southern California to Northern California to enter private practice. His territory is primarily the Southern Salinas Valley where, at the age of 75, he continues to be actively engaged in land surveying and photogrammetry. George has been a member of ACSM since 1946. He is a past chairman of the Southern California Section and presently is on the board for the Northern California Section.

ABSTRACT

The rectangular system of public land surveys over the Public Domain in the United States stipulated the use of initial points for the intersection of principal meridians and baselines. Of these, California has three: Mount Diablo, Humbolt, and San Bernardino. As it happens, over the years, three initial points were established at the San Bernardino Meridian and Baseline. The Washington Monument Expedition was commissioned on June 1, 1966 by the San Bernardino County Museum to make a study of the Initial Points. Why three points were set was the primary question to be answered by the expedition. An important and essential phase of the project was the establishment of Horizontal Control in the area of the Initial Points. This paper covers the planning and organization that pertained to the task of establishing the horizontal control and results thereof. An interesting side effect was the gathering of color slides showing pictorially the terrain and equipment used to conduct the survey.

INTRODUCTION

The task of establishing horizontal control in the vicinity of the Initial Point commenced several weeks prior to the actual expedition which took place over the Labor Day weekend, September 2 through September 5, 1966.

Initially, consideration was given to the positioning by way of resection. There were numerous intersection points in the valley which could be observed, but it was essential that the day of observation selected be one in which the valley was clear of smog. Better accuracy than what resection could give was expected, so after careful reconnaissance in the valley, it was decided to triangulate a new station near the Initial Point using second order or better USC&GS Stations as a base.

THE RESURVEY

USC&GS KELLER 1929, a first order station atop Keller Peak at elevation 7,882 feet and approximately 8½ miles distant, was one of the stations selected to be used. For an azimuth check, USC&GS CRAFT 1929, a first order station atop Butler Peak and five miles from KELLER, was selected. In the valley the USC&GS second order station CRAFTON 1929 was selected primarily because it was a drive station; it's elevation was 3,070 feet and hopefully above the smog, and the triangular scheme afforded a strong figure in which the angle at the new station was about 62°. As an azimuth check on CRAFTON, the USC&GS second order station MENTONE 1929, located within the Lockheed Propulsion Company grounds, and two miles distant from CRAFTON, was chosen. (See Figure 1).

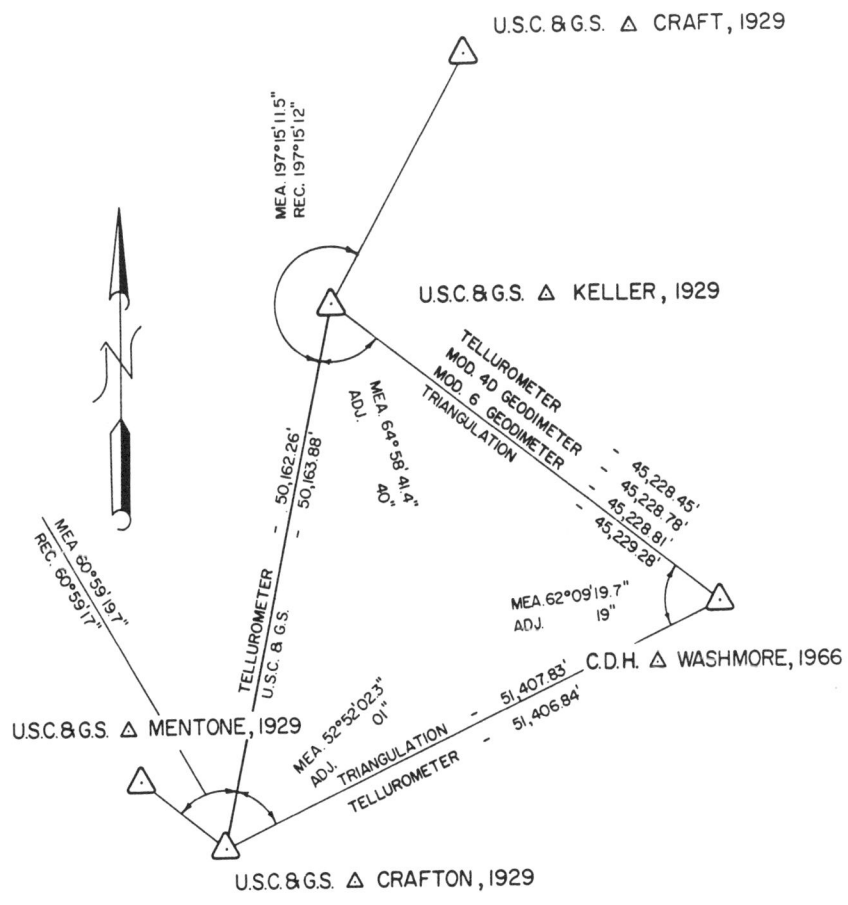

Figure 1. Triangulation and distance measuring plan and results

On August 28,1966, CRAFT 1929 was visited atop Butler Peak just westerly of Big Bear Dam. The lookout tower sits directly over the station. There is about a 9-inch clearance between the station and the floor girder. A light was set over the station aimed towards KELLER. The wiring was arranged so that the towerman could easily energize the light upon notification by radio from the towerman at Keller Peak Lookout.

At CRAFTON, the surface mark was found in a small boulder lying out of position in relation to intact reference marks. The surface mark was salvaged and returned to the USC&GS Field Office in Los Angeles. A temporary point consisting of a "T" Bar with a California Division of Highways cadmium cap set in concrete was established for the station.

Arrangements were completed and copies of the detailed triangulation and distance measuring timetable and routine with sketch were passed out to all participating individuals. The valley routine was under the direction of J.K. Smith of the California Division of Highways, now CALTRANS.

On the morning of September 2,1966, the advance party consisting of Bill Laurie, Desmond Creacy, Bob Eagle, Bill Lawrence, Bill Lawrence Jr., Brent Autler, Richard Norlen, George Gary and four pack horses left the trailhead for Trail Fork Springs. Eagle from Arrowhead Engineering was back-packing their Tellurometer unit. Gary was back-packing, in addition to his personal gear, three buckets containing 21 mirrors which were the reflectors used in conjunction with the Geodimeter shots. The Wild 21-b Tripod was carried by Desmond Creacy in addition to his personal gear. The T-Meter batteries were packed with the horses.

Because of the late start due to trouble in loading the horses in Redlands, the party did not arrive at Trail Fork Springs Camp until 4:30 p.m. After a short rest, Eagle and Gary packed the refelctors, Wild tripod and other gear needed and headed for the Initial Point. The more than three miles to the point was covered in just two hours, and it was almost dark. Radio contact was made with Joe Williams of the California Division of Highways who was at KELLER with his vehicle and the five-watt radio unit loaned to the expedition by the San Bernardino County Surveyor. At the Initial Point, Eagle and Gary had the County two-way portable radio unit. The light tender at CRAFTON, Jerry Imbiorski, of the California Division of Highways, also had a radio unit loaned to the expedition by the County Surveyor.

About forty feet southwesterly of Deputy Surveyor Henry Washington's Initial Point, the California Division of Highways Triangulation Station WASHMORE 1966 was established. The station location was selected for visibility to KELLER and CRAFTON and still be as near to the Initial Point as practicable. A temporary point was set due to darkness. The three buckets of 21 mirrors were set up over the point for distance measuring from KELLER. The mirrors were furnished through the cooperation of Universe Aerial Surveys,Inc., Western Aerial Surveys, and Webb & Carroll. At KELLER, the model 6 Geodimeter measurement was made by

Webb & Carroll with Joe Aguillera as observer and J. Stickman as recorder. The 4-D Geodimeter measurement was made by District 07 (Los Angeles) California Division of Highways. Ralph L. Swingle was observer and Armando Petroni was recorder.

The horizontal angles at KELLER from CRAFTON to WASHMORE to CRAFT were scheduled to be observed by Webb & Carroll immediately after the geodimeter shots. The light at CRAFTON was not visible from KELLER due to heavy smog, so the night's operation was shut down at approximately 11 p.m. Instructions were left with Jerry Imbiorski at CRAFTON that the angle at KELLER would be turned using mirrors at 3:00 p.m. the next day. Eagle and Gary hiked back to camp, arriving at 1 a.m.

After breakfast on Saturday morning, Eagle, Gary and one pack horse left Trail Fork Springs Camp at about 11 a.m. headed for the Initial Point. The horse carried all needed supplies including food, water, and surveying equipment. Upon arriving at the Initial Point, the temporary mark for the California Division of Highways Triangulation Station WASHMORE 1966 was replaced with a standard CDH disk subsurface mark set in an irregular mass of concrete 18" below the surface. The surface mark is a standard CDH disk set in a circular 6" diameter concrete post. The disks are stamped WASHMORE 1966. Two reference marks were set in trees for ease in recovery.

At 3:00 p.m. radio contact was made with J.K. Smith and his son, Jim, at CRAFTON and Joe Williams at KELLER. At 4:00 p.m. Joe Aguillera as observer and J. Stickman as recorder from Webb & Carroll began turning eight positions with their Wild T-2 for the angle at KELLER between WASHMORE and CRAFTON. At CRAFTON, Smith had a heliograph positioned towards KELLER and at WASHMORE a mirror, resting on the Wild tripod, was held over the point. The angle between WASHMORE and CRAFT could not be turned until darkness set in because the light was not visible until then. Again eight positions were turned for that angle.

In the meantime, Bill Young and Hal Robinson, backpacking the theodolite for use at the Initial Point, arrived on the scene. The Wild tripod for the reflectors and the Wild T-2E theodolite were furnished by Surveyor's Service Company of Costa Mesa through Riverside Blueprint Company. Hal and Bill with Dieter Wilken had departed from the trailhead early Saturday morning.

At 8:00 p.m. the three bucket reflectors were set up at WASHMORE for distance measuring from CRAFTON using the 4-B Geodimeter owned by Universe Aerial Surveys and the 4-D Geodimeter owned by Western Aerial Surveys. After repeated attempts by Jack Easton on the model 4-B and Bob Boring on the model 4-D the operation was shut down at 10:00 p.m. No readings could be made because of extreme distance (about 10 miles) and heavy smog. The simultaneous angle turning scheduled for that night at WASHMORE and CRAFTON was postponed until the next day. Eagle, Gary and pack horse arrived at camp at midnight.

On Sunday morning, the entire expedition, except for the camp tender, left camp at 9:00 a.m. with one pack horse, which carried the Tellurometer. Upon arrival at the Initial Point, preparations were made to begin measuring the distance from WASHMORE to CRAFTON and to KELLER using the T-Meter unit. In the meantime, Marvin Kuhn at CRAFTON and Jack Carper at KELLER as master unit were measuring the line CRAFTON and KELLER. At about 1:00 p.m. Bob Eagle used his T-Meter as master unit to measure the distance WASHMORE to KELLER and to CRAFTON. George Gary recorded.

At about 3:00 p.m. the angles at WASHMORE from CRAFTON to KELLER to CRAFT LOOKOUT TOWER were being observed to a heliograph operated by J.K. Smith at CRAFTON and mirror held on tripod at KELLER. By 4:15 p.m. eight acceptable positions were obtained at WASHMORE. This completed the work except for the angle at CRAFTON. Preliminarily, it was planned to set a large sight over WASHMORE to observe from the valley early some clear morning during the coming week.

That evening at Trail Fork Springs Camp, a steak dinner was waiting for the men, and it certainly did taste good after eating cold trail lunches.

On Monday morning, Eagle, Gary, and Lawrence Jr., taking one horse, packed personal gear, reflectors, and Tellurometer and headed for the Initial Point at about 8:00 a.m. The plan was to set a sight over the station WASHMORE and then, instead of looping back to camp, return to Camp Angelus trail by way of Limber Pines Camp and Manzanita Springs. This was done and they arrived at Camp Angelus at 2:30 p.m. The horse was loaded into the trailer parked at the Division of Highways Camp Angelus Maintenance Station and the rest of the party was met at the trail head from Trail Fork Springs. Timing could not have been better. The one horse was unloaded, two horses from Redlands were loaded and they were on their way to the valley floor about 4:00 p.m.

In order to close the angles in the triangle, it was decided to complete that phase of the work on Saturday, September 10 1966. J.K. Smith and Gary packed in with a heliograph via the trail from Camp Angelus. They left the trail head at 6:00 a.m. and arrived at the Initial Point at noon. The heliograph was readied at 12:15 and by 12:45 the angles at CRAFTON were commenced. Kenneth Simpson observed and Jerry Ivy recorded. Both men were from the California Division of Highways and their instrument was a Kern DKM-2 Theodolite. An umbrella was used. Joe Williams and brother Walt were manning KELLER; they too had a heliograph setup. No radios were used because mirror signals had been prearranged by Smith.

About 2:45 p.m. the signal was received at WASHMORE that the angles were completed. WASHMORE signalled in return and in a few minutes were on the way back to the trail head arriving at 6:15 p.m. At 7:00 p.m. they were back in San Bernardino.

The angular closure in the triangle was 2.9 seconds after spherical excess was accounted for. The plane coordinate positioning of the triangulation station and Initial Point can be considered to be second order accuracy. The work was submitted to the USC&GS for final analysis and adjustment. The data was accepted and then published in their standard form. Our preliminary computations are shown in Figure 1.

Monetary contributions to the project are worthy of mention. The Civil Engineers and Licensed Land Surveyors Association of Riverside and San Bernardino Counties donated $100., Art Sullivan of Arrowhead Engineering donated $70., and the Southern California Section of the American Congress on Surveying and Mapping donated $50..

REFERENCES

Laurie, B.D. 1966 Washington Monument Resurvey Expedition: Surveying and Mapping, Vol. XXVI, No. 4, pp.685-688

Laurie, B.D. 1967 Washington Monument Resurvey Expedition: Surveying and Mapping, Vol. XXVII, No. 2, pp.301-313

Laurie, B.D. 1967 Washington Monument Resurvey Expedition: San Bernardino County Museum Association, Vol. XIV, 3 & 4

LEVELING WITHOUT A LEVEL INSTRUMENT
(A TRIGONOMETRIC METHOD)

Charles C. Glover
National Geodetic Survey
Charting and Geodetic Services
National Ocean Service, NOAA
Rockville, Maryland 20852

ABSTRACT

The advent of total station surveying has integrated all survey activities into one instrument, including levels. However, leveling with a total station must be considered unproven to date. Tests performed by the National Geodetic Survey (NGS) and the South Carolina Geodetic Survey have not yet been conclusive enough to determine adequate specifications. (Analysis of the South Carolina data is not yet completed.) Reports from the private sector have been very positive in that third- and even second-order, class II leveling can be accomplished. Leveling with a total station instrument is achieved by measuring a vertical angle (VA) and slope distance (SL/D), then trigonometrically computing the difference of elevation as opposed to a direct measurement as in conventional leveling. The accuracy of trig-leveling is determined primarily by the ability of the total station to resolve the two measured values (i.e., VA and SL/D). Testing, thus far, has also revealed that atmospheric conditions and sight length seem to influence the results. However, sight lengths from 200 to 300 meters appear possible. More testing should reveal optimum sight lengths depending on the instrumentation and atmospheric conditions. If trig-leveling can be proven reliable, it will greatly enhance the total station method for today's surveying. More testing will be done at NGS during the first half of 1991, with the hope of determining realistic trig-leveling specifications.

Developing Transverse Mercator Projection Tables for NAD'83

Joshua S. Greenfeld
Department of Civil & Environmental Engineering
New Jersey Institute of Technology
Newark NJ, 07102

Abstract

Many surveyors who needed to work with the state plane coordinate system from NAD'27 have used projection tables published by USCGS and later by NGS. Recently, NGS has published projection tables for NAD'83 but only for the Lambert Conformal Conic (LCC) projection. Surveyors in states (such as New Jersey) who need to use the Transverse Mercator (TM) projection have expressed a strong interest in having such projection tables at their disposal.

This paper discusses the need for such projection tables, the alternative ways to derive them and some practical usage considerations. An example of TM tables for NAD'83 which have been developed for the state of New Jersey is presented and discussed.

INTRODUCTION

NGS (National Geodetic Survey, NOAA) has completed the definition and computation of the new North American Datum named NAD'83. Consequently, many states revised their official survey base statute from NAD'27 to NAD'83. Many surveyors who were affected by this act found themselves without a simple straight forward method for computing the conversions between geodetic coordinates and plane coordinates, meridian convergence and grid scale factor on NAD'83. For NAD'27 there were projection tables readily available from the US Coast and Geodetic Survey (USCGS) and later on from NGS. NGS is marketing software for these computations on NAD'83, however, there also is a need for projection tables that can be easily used and handled with a simple scientific calculator.

This paper discusses the need for such projection tables, the alternative ways to derive them and some practical usage considerations. An example of TM tables for NAD'83 which have been developed for the state of New Jersey is presented and discussed.

THE NEED FOR TABLES

Before discussing how to develop projection tables, it is fundamental to understand why we need to use such tables in the 1990's. Nowadays, computers are readily available their prices are rather affordable and the necessity for computers in a surveying outfit can hardly be challenged. So why do we need to use tables for our computations rather than purchasing a low cost software which is capable of performing these computations rapidly and effortlessly? One obvious answer to this question is that there are still surveyors who do not use computers, for whatever reasons, or who have a computer for which there is no software for these computations. These surveyors must resort to tables to perform coordinate conversions, meridian convergence and grid scale factor computations. Another way to explain the need for projection tables is to examine what type of computations surveyors do with state plane coordinates and under what circumstances they might need to perform these computations.

When assessing the main use of state plane coordinates systems among several New Jersey surveyors, it was found that the overwhelming majority of these surveyors perform only meridian convergence and grid scale factor computations. These computations are needed to compute closures of connecting traverses that are bases on control points that have coordinate values in the state plane coordinate system. Coordinate conversions are done very rarely as control points descriptions obtained from NGS or DOT contain both geodetic and plane coordinate values. Since grid scale factor and meridian convergence can be computed fairly easily from tables, for a limited number of points a computer program does not introduce a meaningful saving of time and effort.

Another reason for using tables rather than software for plane coordinate computations is to have computation means for those circumstances in which computers are out of reach. For example, sometimes there is a need to perform computations in the field where there is no access to a computer. Another example is the frustrating and unfortunately familiar event when the computer is down and there is an urgent need for such computations. For these and other circumstances, tables are the logical solution.

TRANSVERSE MERCATOR TABLES

In the past, two different types of tables for the Transverse Mercator (TM) projection have been published. The first one by the US coast and Geodetic Survey (USCGS) for NAD'27 and the other one by the US army for the Universal Transverse Mercator (UTM). TM tables for NAD'27 have been developed experimentally (Dracup 1990) and computation is performed by a product of several coefficients. For example (USCGS 1954) the conversion from geodetic to plane coordinates are computed by:

$$x = H \cdot \Delta\lambda" \pm a \cdot b + x_0$$
$$y = y_0 + V \cdot (\frac{\Delta\lambda"}{100})^2 \pm c$$

Where:
- x, y - Plane Easting and Northing
- $\Delta\lambda"$ - Arc distance from the central meridian in "
- x_0 - Grid false Easting
- y_0, H, V, a - Tabulated coefficients as a function of φ
- b, c - Tabulated coefficients as a function of $\Delta\lambda"$

Computations with UTM tables are done by a (odd or even) polynomial for which the coefficients are tabulated. For example (DoA 1964) the conversion from geodetic to plane coordinates are computed by:

$$N = (I) + (II)p^2 + (III)p^2 + A_6$$
$$E = E_0 \pm [(IV)p + (V)p^3 + B_5]$$

Where:
$$p = 0.0001(\Delta\lambda")$$
E, N - Plane Easting and Northing
E_0 - Grid false Easting
$(I) - (V), A_6, B_5$ - Tabulated coefficients as a function of φ

Working with tables developed for NAD'27 is somewhat easier. They use less coefficients and require a smaller number of computations. However, they require a tremendous amount of trial and error effort to derive these coefficients. A polynomial type of tables (such as UTM tables) is easier to derive but they result in a slight increase in computation effort.

In the next section we present our choice for the general form of our projection tables which is a modification of the UTM tables.

DEVELOPING TM TABLES FOR NAD'83

General

Several important factors should be considered when designing and developing computation tables. The most important one is that they must be very simple to use. If the tables are complicated and require many interpolations they defy the rationale behind developing tables as a means for simplifying complex computations. Another important consideration is to design the tables according to their typical use. For example, if the inverse conversion is being computed and thus, we have Easting Northing (E,N), the coefficients for computing (φ, λ) should be given as a function of (E,N). In other words, only simple interpolations should be required avoiding the somewhat more confusing backward interpolations. This kind of tabulation results in a more simple and intuitive computation procedure.

In the past, conversion tables have been derived to meet manual computation (paper and pencil, LOG tables or adding machines) needs. One such need was to avoid computations that involved Degrees, Minutes and Seconds (DMS). Consequently, the tables were designed to work mainly with seconds of arc rather than the whole value of the latitude and longitude. Today, with the advent of hand held calculators, a more natural form of arc units is the Decimal Degree (DD). For example, if a point has a longitude of $73°13'55.12345"$ and the central meridian is $74°30'$, using seconds of arc we have $\Delta\lambda" = -4564.87655"$. It is simpler to key in to the calculator the longitude, convert it to DD and subtract 74.5 to get $\Delta\lambda^{dd}$. Any simple scientific calculator has a built-in function to convert DD to DMS and visa versa. Thus, it makes sense to design the tables for Decimal Degrees rather than for seconds of arc.

Mapping equations for TM on NAD'83

The most comprehensive source of information and equation regarding mapping equations for NAD'83 is presented in the manual "state plane coordinate system of 1983" (Stem 1989). This manual presents the constants for each mapping zone, the mapping equations as well as some suggested forms of use. There are four sets of equations for each projection used in the United States. The first set is for solving the direct conversions, namely, computing state plane coordinates (Northing and Easting) from geodetic coordinates (Latitude and Longitude). The second set of equations is used to compute geodetic coordinates from plane coordinates. This set of equations is sometimes referred to as the inverse conversion. The third and fourth sets of equations are used for computing the meridian convergence and the grid scale factor respectively. Actually, for the meridian convergence and the grid scale factor there are two sets of equations for each. One set assumes that we are given geodetic coordinates while the other assumes that we have state plane coordinates. These equations for the Transverse Mercator have, in general, a form of a polynomial for which several coefficients have to be computed.

For example, the mapping equations for the direct conversion are:

$$N = S - S_0 + N_0 + A_2 L^2 [1 + L^2(A_4 + A_6 L^2)]$$
$$E = E_0 + A_1 L[1 + L^2(A_3 + L^2(A_5 + A_7 L^2))]$$

Where:
$$L = \frac{\lambda - \lambda_0}{\cos \varphi}$$

- E, N - Plane Easting and Northing
- φ, λ - Geodetic Latitude and Longitude
- E_0, N_0 - Grid false Easting and Northing, respectively
- S_0, λ_0 - Constants for meridional distance and central meridian, respectively
- $A_1 - A_7, S$ - Coefficients as a function of the mapping constants and φ

The linear units in these equations are in meters while the arc units are in radians. A similar form of a polynomial is used for the inverse conversion, meridian convergence and grid scale factor computations.

Mapping Equations using tabulated coefficients

Based on the above discussions and on a modified version of the equation published by NGS (Stem 1989) the following mapping equations have been derived:

1) **Direct Conversion.** Solving (N,E) from (φ, λ)

$$L = \lambda^d - \lambda_0^d$$

$N = S' + A_2 \cdot L^2 + A_4 \cdot L^4$ - *2nd order correction
$E = E_0 + A_1 \cdot L + A_3 \cdot L^3$ - *2nd order correction

Where:

- λ^d - Geodetic Longitude in $D^\circ.ddddd$ (Decimal Degrees)
- λ_0 - Geodetic Longitude of the Central Meridian in DD.
- E_0 - Grid false Easting (m)
- S', A_2, A_4, A_1, A_3 - Tabulated coefficients as a function of φ
- φ - Geodetic Latitude in D° MM' SS.SSSS"
- N, E - Grid Northing and Easting (m)

2) **Inverse Conversion.** Solving (φ, λ) from (N,E)

$$Q = \frac{E - E_0}{1,000,000}$$

$$\varphi = \varphi' + B_2 \cdot Q^2 + B_4 \cdot Q^4 \text{ - *2nd order correction}$$
$$\lambda = \lambda_0 + B_1 \cdot Q + B_3 \cdot Q^3 + B_5 \cdot Q^5 \text{ - *2nd order correction}$$

Where:

- λ - Geodetic Longitude in $D^\circ.ddddd$ (Decimal Degrees)
- φ - Geodetic Latitude in $D^\circ.ddddd$ (DD)
- λ_0 - Geodetic Longitude of C.M. in DD
- φ' - Footpoint Latitude. Tabulated as a function of N.
- E_0 - Grid false Easting (m)
- E - Grid Easting (m)
- B_2, B_4, B_1, B_3, B_5 - Tabulated coefficients as a function of N

* 2nd order corrections are tabulated and are necessary only for accuracies better than ±0.001mm or ±0.0001".

3) **Meridian Convergence** γ

$$\gamma = D_1 \cdot Q + D_3 \cdot Q^3$$

Where:

- γ - Meridian Convergence in $D^\circ.ddddd$ (Decimal Degrees)
- $Q = \frac{E - E_0}{1,000,000}$ - (same as in inverse conversion)
- E_0 - Grid False Easting (m)
- E - Grid Easting (m)
- D_1, D_3 - Tabulated coefficients as a function of N

4) Grid (Point) Scale Factor K

$$\boxed{K = K_0 + G_2 \cdot Q^2}$$

Where:

$Q = \frac{E - E_0}{1,000,000}$ - (same as in inverse conversion)
K_0 - Grid S.F. assigned to the Central Meridian
E_0 - Grid false Easting (m)
E - Grid Easting (m)
G_2 - Tabulated coefficient as a function of N

It should be noted that even though the coefficients in these equations bear the same notation (A,B,D and G) as in NGS's publication (Stem 1989), there is no one to one correspondence between them. The modifications include conversion to Decimal Degrees, aggregation of several coefficients into one single coefficient and multiplication of these coefficients by variables which are a function of the latitude (φ).

The second order correction for the direct and inverse conversions is needed because the approximation of a linear interpolation between two tabulated values is not sufficient for providing the highest accuracies. They were computed by observing the differences between the interpolated values and those computed from the rigorous equations. It should be noted that for most practical purposes (accuracies of up to 0.005 ft) there is no need for considering these corrections.

IMPLEMENTATION AND EXAMPLES

Projection tables for Transverse Mercator have been developed for the state of New Jersey. The coefficients for the direct conversion are tabulated as a function of the given latitude (φ) while the coefficients for the inverse conversion are tabulated as a function of the given northing (N). The coefficients for the meridian convergence and the grid scale factor are tabulated as a function of the northing only. The reason for this is that surveyors use almost exclusively plane coordinates in traversing and other traditional surveying control work. Rarely, if ever, do they use geodetic coordinates in their computations. To work with geodetic coordinates they would need to use special software in which case the use of tables is obsolete anyway. As mentioned earlier, NGS or DoT's geodetic control information include both state plane and geodetic coordinates as well as meridian convergence and grid scale factor. Thus, for traditional geodetic control work, the tabulation of meridian convergence and grid scale factor as a function of state plane coordinates only is sufficient.

In addition to tables, special computation forms have been designed. The importance of the forms is twofold. Firstly, it shows the computation procedure in a more illustrative fashion. It is easier to understand and to follow the progression of the computation when one is following a well defined procedure. Secondly, a form provides a neat record for future references or for computation checks in case of an error.

φ	S'	A_1	A_2	A_3	A_4
38°50'	0.0000	-86 820.579 81	475.091 38	-0.952 45	0.032 12
$\Delta/''$	30.8337 12	0.337 5115	0.001 0133	0.000 0457	-0.000 0003
38°51'	1 850.0227	-86 800.329 12	475.152 18	-0.949 71	0.032 10
$\Delta/''$	30.8337 98	0.337 6345	0.001 0108	0.000 0457	-0.000 0003
38°52'	3 700.0506	-86 780.071 05	475.212 83	-0.946 97	0.032 08
$\Delta/''$	30.8338 88	0.337 7573	0.001 0080	0.000 0455	-0.000 0002
38°53'	5 550.0839	-86 759.805 61	475.273 31	-0.944 24	0.032 07
$\Delta/''$	30.8339 77	0.337 8802	0.001 0053	0.000 0457	-0.000 0003
38°54'	7 400.1225	-86 739.532 80	475.333 63	-0.941 50	0.032 05
$\Delta/''$	30.8340 63	0.338 0030	0.001 0028	0.000 0455	-0.000 0003
38°55'	9 250.1663	-86 719.252 62	475.393 80	-0.938 77	0.032 03
$\Delta/''$	30.8341 53	0.338 1257	0.001 0000	0.000 0455	-0.000 0002
38°56'	11 100.2155	-86 698.965 08	475.453 80	-0.936 04	0.032 02
$\Delta/''$	30.8342 40	0.338 2485	0.000 9973	0.000 0457	-0.000 0003
38°57'	12 950.2699	-86 678.670 17	475.513 64	-0.933 30	0.032 00
$\Delta/''$	30.8343 30	0.338 3712	0.000 9947	0.000 0455	-0.000 0003
38°58'	14 800.3297	-86 658.367 90	475.573 32	-0.930 57	0.031 98
$\Delta/''$	30.8344 17	0.338 4940	0.000 9920	0.000 0455	-0.000 0002
38°59'	16 650.3947	-86 638.058 26	475.632 84	-0.927 84	0.031 97
$\Delta/''$	30.8345 05	0.338 6165	0.000 9893	0.000 0455	-0.000 0003
39°00'	18 500.4650	-86 617.741 27	475.692 20	-0.925 11	0.031 95
$\Delta/''$	30.8345 95	0.338 7392	0.000 9867	0.000 0455	-0.000 0003

Table 1. Tabulated Coefficients for the direct conversion from geodetic coordinates to New Jersey's state plane coordinates.

N	φ'	B_1	B_2	B_3	B_4	B_5
0.	38.8333 33333	-11.5180 0647	-0.567 8114	0.1082 73	0.00803	-0.0019
Δ/m	0.0000 09008 8975	-0.0000 0145 2560	-0.000 0001 8165	0.0000 000530		
2 000.	38.8513 51128	-11.5209 1159	-0.568 1747	0.1083 79	0.00804	-0.0019
Δ/m	0.0000 09008 8700	-0.0000 0145 3855	-0.000 0001 8175	0.0000 000530		
4 000.	38.8693 68868	-11.5238 1930	-0.568 5382	0.1084 85	0.00805	-0.0019
Δ/m	0.0000 09008 8415	-0.0000 0145 5160	-0.000 0001 8185	0.0000 000530		
6 000.	38.8873 86551	-11.5267 2962	-0.568 9019	0.1085 91	0.00806	-0.0019
Δ/m	0.0000 09008 8140	-0.0000 0145 6465	-0.000 0001 8190	0.0000 000530		
8 000.	38.9054 04179	-11.5296 4255	-0.569 2657	0.1086 97	0.00807	-0.0019
Δ/m	0.0000 09008 7860	-0.0000 0145 7765	-0.000 0001 8200	0.0000 000530		
10 000.	38.9234 21751	-11.5325 5808	-0.569 6297	0.1088 03	0.00807	-0.0019
Δ/m	0.0000 09008 7585	-0.0000 0145 9075	-0.000 0001 8210	0.0000 000530		
12 000.	38.9414 39268	-11.5354 7623	-0.569 9939	0.1089 09	0.00808	-0.0019
Δ/m	0.0000 09008 7300	-0.0000 0146 0380	-0.000 0001 8220	0.0000 000535		
14 000.	38.9594 56728	-11.5383 9699	-0.570 3583	0.1090 16	0.00809	-0.0019
Δ/m	0.0000 09008 7025	-0.0000 0146 1690	-0.000 0001 8230	0.0000 000535		
16 000.	38.9774 74133	-11.5413 2037	-0.570 7229	0.1091 23	0.00810	-0.0019
Δ/m	0.0000 09008 6745	-0.0000 0146 2995	-0.000 0001 8240	0.0000 000530		
18 000.	38.9954 91482	-11.5442 4636	-0.571 0877	0.1092 29	0.00811	-0.0019
Δ/m	0.0000 09008 6465	-0.0000 0146 4310	-0.000 0001 8245	0.0000 000535		
20 000.	39.0135 08775	-11.5471 7498	-0.571 4526	0.1093 36	0.00811	-0.0019
Δ/m	0.0000 09008 6190	-0.0000 0146 5620	-0.000 0001 8255	0.0000 000540		

Table 2. Tabulated Coefficients for the inverse conversion from New Jersey's state plane coordinates to geodetic coordinates.

N	D_1	D_3	G_2
0.	7.2224 4780	-0.097 048	0.012 310
Δ/m	0.0000 0232 4095	-0.000 000 0559	
10 000.	7.2456 8875	-0.097 607	0.012 310
Δ/m	0.0000 0232 9951	-0.000 000 0564	
20 000.	7.2689 8826	-0.098 171	0.012 309
Δ/m	0.0000 0233 5841	-0.000 000 0567	
30 000.	7.2923 4667	-0.098 738	0.012 309
Δ/m	0.0000 0234 1766	-0.000 000 0571	
40 000.	7.3157 6433	-0.099 309	0.012 309
Δ/m	0.0000 0234 7724	-0.000 000 0575	
50 000.	7.3392 4157	-0.099 884	0.012 309
Δ/m	0.0000 0235 3718	-0.000 000 0579	

Table 3. Tabulated Coefficients for computing the meridian convergence and the grid scale factor.

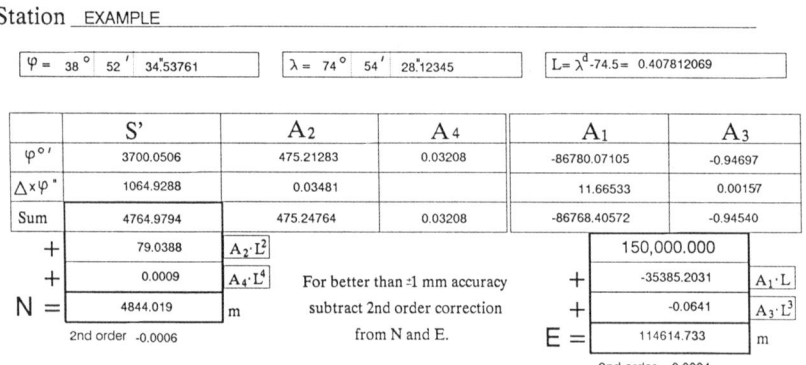

Table 4. Form for computing New Jersey's state plane coordinates from geodetic coordinates

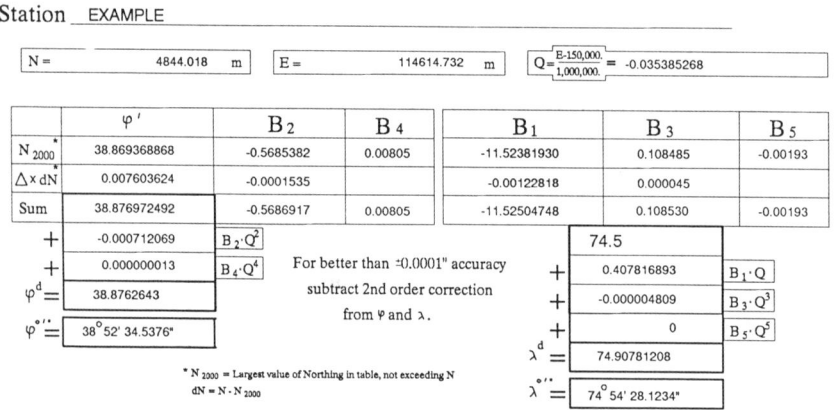

Table 5. Form for computing geodetic coordinates from New Jersey's state plane coordinate system.

Station ____EXAMPLE____

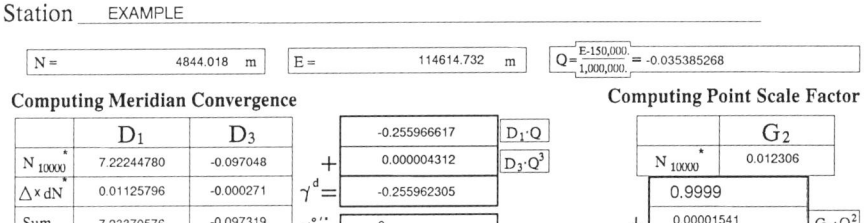

	Computing Meridian Convergence				
	D_1	D_3	-0.255966617	$D_1 \cdot Q$	
N^*_{10000}	7.22244780	-0.097048	0.000004312	$D_3 \cdot Q^3$	
$\Delta \times dN$	0.01125796	-0.000271	$\gamma^d =$ -0.255962305		
Sum	7.23370576	-0.097319	$\gamma^{\circ\,'\,''} =$ -0°15' 21.46"		

* N^*_{10000} = Largest value of Northing in table, not exceeding N
dN = N - N^*_{10000}

Computing Point Scale Factor		
	G_2	
N^*_{10000}	0.012306	
	0.9999	
+	0.00001541	$G_2 \cdot Q^2$
K =	0.99991541	

Table 6. Forms for computing meridian convergence and grid scale factor

Second Order Correction for Northing (N) in Meters

	Longitude					
$\Delta\varphi"$	73°30'	74°00'	74°30'	75°00'	75°30'	75°45'
10	0.0003	0.0003	0.0003	0.0003	0.0003	0.0003
20	0.0005	0.0005	0.0005	0.0005	0.0005	0.0005
30	0.0006	0.0006	0.0006	0.0006	0.0006	0.0006
40	0.0005	0.0005	0.0005	0.0005	0.0005	0.0005
50	0.0003	0.0003	0.0003	0.0003	0.0003	0.0003

Second Order Correction for Easting (E) in Meters

	Longitude					
$\Delta\varphi"$	73°30'	74°00'	74°30'	75°00'	75°30'	75°45'
10	0.0005	0.0003	0.0000	0.0003	0.0005	0.0006
20	0.0008	0.0004	0.0000	0.0003	0.0007	0.0009
30	0.0009	0.0004	0.0000	0.0004	0.0008	0.0010
40	0.0008	0.0004	0.0000	0.0003	0.0007	0.0009
50	0.0005	0.0003	0.0000	0.0003	0.0005	0.0006

Second Order Correction for Latitude (φ) in Seconds of Arc

	Easting					
ΔN_m	0	50,000	100,000	150,000	200,000	250,000
500	-0.00002	-0.00002	-0.00002	-0.00002	-0.00002	-0.00002
1000	-0.00003	-0.00003	-0.00003	-0.00003	-0.00003	-0.00003
1500	-0.00002	-0.00002	-0.00002	-0.00002	-0.00002	-0.00002

Second Order Correction for Longitude (λ) in Seconds of arc

	Easting					
ΔN_m	0	50,000	100,000	150,000	200,000	250,000
200	0.00007	0.00005	0.00002	0.00000	-0.00002	-0.00005
400	0.00012	0.00008	0.00004	0.00000	-0.00004	-0.00008
600	0.00016	0.00011	0.00005	0.00000	-0.00005	-0.00011
800	0.00018	0.00012	0.00006	0.00000	-0.00005	-0.00012
1000	0.00019	0.00013	0.00007	0.00000	-0.00007	-0.00013
1200	0.00018	0.00012	0.00006	0.00000	-0.00005	-0.00012
1400	0.00016	0.00011	0.00005	0.00000	-0.00005	-0.00011
1600	0.00012	0.00008	0.00004	0.00000	-0.00004	-0.00008
1800	0.00007	0.00005	0.00002	0.00000	-0.00002	-0.00005

Table 7. Second order corrections for results better than ±0.001 mm or ±0.0001" (seconds of arc)

Tables 1-3 show the coefficient for the direct conversion, inverse conversion and the coefficient for the meridian convergence and grid scale factor, respectively. Tables 4-6 show the computation forms and an example for these computations. The second order corrections for more accurate computations are presented in Table 7.

SUMMARY

As a result of the completion of NAD'83, surveyors found themselves with a need for a simple and straight forward means for performing computations on this new datum. Projection tables are one important alternative in developing tools for alleviating the computation burden. Even in the 1990's there is still a need for projection tables for those surveyors who either do not have computers or are under circumstances which do not allow access to computers.

Projection tables have been developed for the Transverse Mercator projection. The tabulated coefficients have been computed based on mapping equations published by NGS (Stem 1989). The computations of the conversions from state plane coordinates to geodetic coordinates, geodetic coordinates to state plane coordinates, meridian convergence and grid scale factor, are carried out using a polynomial for which the coefficients are tabulated. The main objective in developing these tables was to make them suitable for hand held calculators and to make them as easy to work with as possible. In addition to tables, computation forms have been developed as well.

REFERENCE

DoA, TM 5-237, Surveying computation manual, Department of the army technical manual, October, 1964.

USCGS, Publication No. 316, Plane Coordinate Projection Tables New Jersey, USCGS, 1954.

Dracup F. Joseph, Personal communication, ACSM annual convention, Denver CO., 1990.

Stem E. James, State Plane Coordinate System of 1983, NOAA Manual NOS NGS 5, 1989

GEODETIC SURVEYING IN SOUTHWESTERN PENNSYLVANIA
John Hamilton
Geodetic Engineer
Hartmann Associates
5637 Brownsville Rd.
Pittsburgh, PA 15236

ABSTRACT

As GPS surveying becomes more common, the use of geodetic control will increase. Historically, geodetic control has been performed mostly by government agencies, and is therefore not familiar to many practicing surveyors. This paper addresses the experiences gained over three years working in a six county region of southwestern Pennsylvania. The history of geodetic control in the region is first discussed, followed by a discussion of select projects. Much emphasis is placed on vertical control, and results obtained with GPS and precise levels.

INTRODUCTION

Hartmann Associates, Inc. is a surveying firm located in a southern suburb of Pittsburgh, PA. They have been actively involved in GPS surveying since early 1988. The firm participated with Geo/Hydro in the first GPS survey of an urban area in 1984. Over the past three years, a network has been built up through various projects. Much experience has been gained dealing with control of different accuracies and origins. A brief history of control in the region will be given along with a description of recent work. An analysis of results will also be presented which shows the advantages and disadvantages of using existing control of different epochs.

Description of Area

Southwestern Pennsylvania is located on the Allegheny Plateau, which is on the western front of the Appalachian Mountains. While termed a plateau, it does not appear to be one to a casual observer, since it is deeply eroded by the rivers and streams. This resulted in very steep-sided valleys, with the only flat areas being on the flood plains of the streams. The remains of the plateau (now seen as flat hilltops) are located at about 1,200 feet above MSL, and the steep sided valleys are cut 400-600 feet below. The entire area is underlain by various coal seams, the most famous being the Pittsburgh Coal. Much of the coal in Allegheny County surrounding Pittsburgh has been mined, and subsidence is a major problem.

One consequence of being located on a plateau is that the underlying rocks are nearly horizontal, with small local anticlines and synclines. This results in relatively smooth geoid. However, the large upthrust ridges to the east at the western front of the Applachians do cause a rather abrupt change. This is due to a large amount of excess mass caused by the ridges. It has been estimated that the Appalachians once reached an altitude of 14,000 feet, higher than much of the Rocky Mountains, before being eroded to their present 2000-3000 feet. For example, typically the deflections range from 0" to 2" on the plateau, but have values as high as 10" in the mountains. Since Chestnut Ridge is the last mountain until the Rockies, it was a natural location for triangulation stations. Therefore, many of the first order stations in the area are located at the crest of the ridge. Another interesting phenomena caused by the topography is the relative ease with which a survey station can be established which has good visibility to adjacent stations. Unfortunately, this also means that very few existing horizontal control stations have accurate elevations. The vast majority of the primary bench marks are located along the rivers, some 500-600 feet below the hilltops.

EXISTING CONTROL NETWORKS

There have been various control networks established in the area beginning in the 1890´s. These early works were usually for mapping purposes, and were not very accurate. Beginning in the 1920´s, concerted efforts by various agencies began to establish geodetic quality control.

U.S. Geological Survey

There was extensive mapping type surveys done by the USGS dating back 100 years. Much of the initial work was done on the old NAD (also known as U.S. Standard Datum), and was never readjusted. In addition, there are several level lines run by the USGS which were listed by USC&GS in a 1903 report. In the control lists for USGS, there are numerous "isolated bench marks" listed as being established in 1899 and 1903. The USC&GS control lists show various lines run by the USGS in 1889? and 1899. Most of these isolated marks were actually part of these original lines which were later rerun by USC&GS. The more recent activities of the USGS has been the establishment of third order triangulation and level nets. A second order traverse was run in 1961 in the southern corner of the state which NGS publishes in their lists.

U.S. Army Corps of Engineers

Beginning in 1895, the Corps (previously known as the United States Engineers) began surveys to establish a Harbor Line System. The origin was a plate on the Davis Island Lock and Dam, located about 4½ miles downstream from the confluence of the Monongahela and Allegheny Rivers, and was based on the "true" meridian through that point. This system was then extended up the Monongahela and Allegheny Rivers by triangulation along both banks. This origin was indirectly tied in 1989 by this firm. No data remains from the harbor line survey other than maps showing the location of stations and the bearing and distance between them. Therefore, the procedures used and adjustment performed are not known. Levels were also run up each river from Pittsburgh. Of significance is a level line classified as second order which runs along the only section of the Monongahela River not covered by USC&GS. This line, run in 1941, controlled several USGS lines near Masontown, and will be discussed later. The above mentioned 1903 USC&GS report mentions a line run by the U.S. Engineers from BM P.R.R. 100 at Pittsburgh along the Ohio River to Kentucky.

City of Pittsburgh

A report was issued in 1911 titled "Improvements Necessary to Meet the City´s Present and Future Needs" by Frederick Law Olmstead . This report stated "No city of equal size in

America, or perhaps in the world, is compelled to adapt its growth to such a difficult complication of high ridges, deep valleys, and precipitious slopes, as Pittsburgh." Quoting from his recommendations: "An accurate framework of reference points needs to be established, including: 1) the gradual systematic setting of permanent street monuments throughout the city to serve as reference points for the definite determination of street locations and for all public and private local surveys; 2) the accurate determination of the locations and elevations of these and other monuments and bench marks in reference to a single general system of coordinates and in reference to the U.S. Government bench; and 3) as a means of accomplishing these ends, an accurate geodetic triangulation of the district, supplemented by the necessary precise traverse work and precise leveling, all fully checked and compensated for errors."

Because of these recommendations, a primary triangulation network consisting of 103 stations (later expanded to 176 stations) and 11 baselines was established. The intended accuracy of this work was 1:100,000. (Note that first order at this time was 1:25,000). This network was controlled by the 11 baselines and by two observed astronomic azimuths. The report of survey states "geographic position is determined by accepting a mean of the position values of the three United States Geological Survey stations which have been incorporated in the city triangulation". This means the the network was computed on NAD, and the grid system was based on NAD raised to an elevation of 1000 feet. Later, when NAD 1927 became available, the geodetic coordinates were apparently recomputed, and state plane coordinates were published for the primary network.

Precise leveling by the city Bureau of Surveys had begun in 1912, and by 1925 consisted of 435 miles of double run levels over permanent bench marks. This was done with wooden rods, which may have introduced some biases in to the network. Also, the elevations are based on a pre-NGVD 1929 USGS elevation for bench mark P.R.R. 100, located in downtown Pittsburgh. This differs from NGVD 1929 by about 0.28 feet. I have not seen any indication that other bench marks were constrained. Therefore, the 0.28 feet is not consistent. Leveling continued after 1925, but the amount is not known.

Precise traverses were run between triangulation stations for many years. This work resulted in an estimated 18,000 traverse stations, many monumented by disks or pins in concrete inside cast iron boxes. A card file is maintained by the city of these traverse monuments which contain a description with elevation and *unadjusted* coordinates. City subdivision regulations require all subdivision surveys to be tied in to the city grid system. Few users of these monuments are aware that the system is at an elevation of 1000 feet, nor are they aware that the system is based on old NAD. MAny of these stations have an elevation. Curiously, many cards show two elevations, one from the Department of City Planning, and the other from the Department of Public Works. They are usually unadjusted elevations.

The high quality of the precise triangulation suggests that

this system could be readjusted on NAD 1983, benefiting from GPS observations to strengthen the network. This possibility was seriously investigated. It was found that all the field notes and computation records had been destroyed in 1962 in the mistaken belief that photogrammetry had made the data obselete. Therefore, the only remaining data pertaining to this survey are books containing adjusted NAD geodetic and grid coordinates, and adjusted NAD 1927 geodetic and state plane coordinates for the triangulation. Therefore, a true readjustment is not possible, but transformations could be done to mapping accuracy. Several level diagrams with junction points, closures and distance leveled were also found. These could conceivably be used to readjust the level nets in the areas covered by the diagrams.

It is interesting to note the care with which the network was designed and constructed. This work was done under the direction of R. H. Randall, and many of the men who participated in the survey left to work with the U.S. Coast and Geodetic Survey (now National Geodetic Survey). Unfortunately, much of this work faded into obscurity until recently. The 1984 City of Pittsburgh GPS survey resurrected much of this data, as explained later.

National Geodetic Survey
Prior to 1935, all leveling through Pittsburgh had been done by the USGS, Pennsylvania Railroad, and the Baltimore and Ohio Railroad. These lines were all rerun either during 1935 or 1941. The only leveling done after 1941 was for the Interstate Highway program, and a Basic Net A releveling project in 1983. This latter project will be discussed later in the paper. Due to the subsidence mentioned earlier, many of these elevations are no longer valid 50 years or more after being established. In the immediate Pittsburgh area, many of the city bench marks were tied by NGS, and further south many of the harbor line monuments were used.

In 1927, the USC&GS brought a single chain of quadrilaterals through southwestern Pennsylvania, and did additional work in 1933 and 1941. An areal densification consisting of first and second order triangulation was done in the vicinity of Pittsburgh in 1958, about the same time as a second order highway traverse was done along the future routes of I-79 and I-70. A first order triangulation-traverse survey with Geodimeter lengths was done south of Pittsburgh in 1975. Some new stations were added, and old lines were strengthened by the EDM distances. It is important to note that these stations were not adjusted on NAD 1927, but were included in NAD 1983. An important feature of the horizontal control in this area is the Transcontinental Traverse (TCT) work which passes through the region near Uniontown.

1984 CITY OF PITTSBURGH GPS SURVEY
In 1984, the City of Pittsburgh Department of City Planning undertook an effort to create new base mapping. Aerial Data Reduction Associates was retained to perform aerial mapping. ADR subcontracted Hartmann Associates Inc. and Geo/Hydro to provide ground control for the mapping. At this time, the existence of a network of points with state plane coordinates was not known. Some individuals were aware of a tri-

angulation network, but the knowledgeable participants had either long ago retired or died. Despite being a new technology (about 1 year old at that time), GPS was seen as a tremendous time and cost savings over conventional methods. As a result, the Macrometer V-1000 system was used to survey a network of 58 points in the summer of 1984. This is believed to be the first survey of an entire city using GPS. There were 46 new stations, 2 first order and 2 second order horizontal control points, 8 existing city stations (two of which were the first order NGRS stations), and 1 second order bench mark included. Several of the stations were later tied vertically to provide information for orthometric height estimation. This survey was meant to be second order, i.e. 20 ppm, but certain lines were to be first order, 10 ppm. In 1989, the city awarded Hartmann Associates a contract to analyse and update the 1984 work. This will be discussed later.

The coordinates of the above network were published based on the NAD 1927 position of one station, as well as the NGVD 1929 elevations of seven others. An algorithm was computed by Geo/Hydro to convert between city grid and state plane coordinates, based on the city triangulation stations included in the survey.

This project led to a Survey Control System Update project performed by this firm. During this research, much previously "forgotten" information was uncovered. This included field search for all USGS, USC&GS, and City monuments located within or near the city. Out of 859 stations searched for, 359 were recovered. One of the original baselines established for the network was recovered intact. A Topcon DM-S3 was used to measure the distance. The reduced distance differed from the published distance by 0.004 meters, which is 4.5 ppm.

RECENT GEODETIC WORK

Because of the success of the 1984 survey, Hartmann Associates decided to begin offering GPS surveys. In early 1988, three Trimble 4000SL receivers were purchased.

Monongahela Valley Expressway

After three relatively small projects, the first large project to be performed started in the spring of 1988. The Pennsylvania Turnpike Commission was planning a toll road from Morgantown, West Virginia to Pittsburgh. The route was broken into seven sections, and contracts were awarded to mapping consultants to provide 1:2400 maps of the proposed routes. Six of the seven photogrammetric firms subcontracted Hartmann Associates to perform control work. More than 100 photo control points were surveyed with GPS. Since the project was flown before awarding the contracts, the entire job consisted of photo id points. At that time the Turnpike Commission (and Penn DOT) would not accept GPS determined elevations. Three of the six decided to run their own vertical control, while Hartmann Associates performed the differential levels for the other three. We used three wire leveling procedures with an automatic level and standard Philadelphia rod for the first two projects. Weeks were spent trying to resolve small discrepancies found, some due to procedures used and others due to inconsistent bench marks. One major problem was found to be the use of high rods. The level line would start along the river, proceed up large hills to control photo points and then tie in to another BM along the river. The closure would be excellent, but the GPS was indicating differences of as much as 0.22 feet for points along the line. It was determined that there was a systematic bias in the high rod readings, with the bias canceling back out as the line came back down in elevation. The third project, located in the City of Pittsburgh, had most of the id points located on the old river bed going through the heart of Pittsburgh. This long abandoned plain was located some 200-300 feet above the present rivers. Long level runs with significant elevation changes would be required. It was decided to run between second or first order bench marks using a Wild NA2 with parallel plate micrometer and invar rods. FGCC second order specifications for double simultaneous differential leveling were followed. A program was developed for the HP 41CX which verified and stored the data collected. This data was then downloaded to a PC in the office, where another program summarized and output the level run. The first line ran a distance of 3.88 miles and had a closure of 0.006 feet. The second line, with a much larger difference in elevation (385 ft), ran for 5.7 miles and had a closure of 0.030 feet. No manual computations or reductions were required due to the automation introduced. The only notes kept were descriptions of BM's and TBM's established. This method was such a success that all future work of a significant size was done using the method.

The GPS work for this project covered a linear distance of approximately 90 miles. Although the accuracy requirement was for second order class II (50 ppm), all work was done to first order (10 ppm) specifications. Polaris observations were also done at each station established to provide an azimuth reference for future work. Closures were better than first order on NAD 1983, and somewhat less on NAD 1927. One first order station near Pittsburgh was found to have moved horizontally approximately 0.75 feet due to mine subsidence. Two large water tanks had been constructed nearby a year after the last USC&GS occupation. The distances and directions to the reference marks checked, so it is believed that the entire area around the monument subsided as a unit.

Monongahela River Survey

After numerous small projects in the region, the next large project was done for the Pittsburgh District of the U. S. Army Corps of Engineers. The Corps decided to remap the entire Monongahela River Valley from Pittsburgh to Fairmont, Wv, a distance of 130 river miles. We proposed setting intervisible pairs of permanent monuments approximately every five river miles, and to perform precise differential levels over the entire length. The intended accuracy of the GPS network was to be 2 ppm, and the levels were to be run to FGCC second order class I specifications. This proposal was submitted to the Engineer Topographic Laboratory for review and was approved.

Fourteen first order NGRS stations (including 2 TCT stations) were used to control the GPS survey. One station, located on the roof of the Engineering Building at West Virginia University, did not fit the rest of the network. Since the azimuth mark was a second order station, the azimuth and distance to it were field measured. This had a misclosure of 0.015 meters, thereby proving that the main station had been displaced about 0.6 m. Two second order traverse stations were included, but were not constrained. To check the internal accuracy of the network, loop closures were checked. A loop is defined here as a closed figure containing baselines from at least two different sessions. The following is a table summarizing the results:

# of lines	Distance	Closure	Precision	Vert. closure
5	63022 m	0.066 m	1.05 ppm	-0.022 m
4	31940 m	0.058 m	1.82 ppm	-0.048 m
11	97532 m	0.102 m	1.05 ppm	0.089 m
4	40433 m	0.017 m	0.43 ppm	-0.000 m
6	31419 m	0.051 m	1.62 ppm	-0.005 m
5	54884 m	0.035 m	0.63 ppm	-0.020 m
8	78359 m	0.098 m	1.25 ppm	0.062 m
7	64197 m	0.090 m	1.40 ppm	-0.015 m
5	70251 m	0.074 m	1.05 ppm	-0.058 m
14	125865 m	0.130 m	1.04 ppm	-0.057 m
9	84010 m	0.054 m	0.77 ppm	-0.045 m

As mentioned above, the intended accuracy for this network was 2 ppm. It clearly exceeded that requirement. These loops cover the entire project area. Closures between the control stations were also checked. The average closure was 3.48 ppm, and ranged from 0.54 ppm to 7.94 ppm for the 31 closures checked (the latter was over a short line, 4.3 km). Several interesting closures should be mentioned. The closure between the two TCT stations was 3.28 ppm over 22.3 km. One of the lines directly measured in the 1975 NGS survey and the 1989 survey had an agreement of 0.54 ppm over 32.3 km. The closure between the farthest north station and the farthest south station was 0.163 m over 183 km, for a precision of 0.89 ppm.

Another analysis can be done by holding one station in the network fixed (TCT station SUMMIT AMS), and computing the misclosure at the other stations on NAD 1983. Note that the adjustment took into account the ellipsoidal height of each station as computed by GEOID4 from NGS.

Station Name	Dist. to SUMMIT	Azimuth	Dist.	Precision
DUNBAR 1927	4303 m	259°	0.032 m	7.44 ppm
HALL TPC 1969	22076 m	179°	0.046 m	2.07 ppm
ET1 DCB 1961*	19016 m	300°	0.237 m	12.46 ppm
GILES 1975	42220 m	22°	0.074 m	1.75 ppm
MALONE 1975	45989 m	56°	0.068 m	1.48 ppm
STENTZ 1975	28765 m	271°	0.060 m	2.09 ppm
DIXIE 1941	75301 m	8°	0.206 m	2.74 ppm
CHESTNUT 1975	19420 m	343°	0.114 m	5.87 ppm
SPEERS 1961*	35180 m	17°	0.202 m	5.74 ppm
SHANNON 1927	63769 m	35°	0.114 m	1.79 ppm
FAUST ACPC 1927	52890 m	2°	0.140 m	2.65 ppm
PERRY 2 1958	83504 m	70°	0.036 m	0.43 ppm
HILLSBORO 1927	36204 m	2°	0.143 m	3.95 ppm
SHEPPLER 1927	44158 m	24°	0.091 m	2.06 ppm

*denotes second order traverse station

As mentioned earlier, the 1975 stations were not computed on

NAD 1927. There were 23 closures checked on NAD 1927, with the average being 22.9 ppm. The low was 2.35 ppm, and the high was 118.58 ppm (once again over the 4.3 km line). The TCT line had a closure of 25.99 ppm. The closure between the two extreme stations mentioned above was 3.564 m over the distance of 182943 m, giving a precision of 19.48 ppm.

The excellent job done by the NGS on NAD 1983 is clearly shown by this analysis. It can be said that the first order network in this area has an accuracy of 20 ppm at best on NAD 1927. It easily meets the claimed accuracy on NAD 1983, namely 10 ppm.

The vertical control network started just downstream of the junction of the Monongahela and Allegheny Rivers and followed the Monongahela River upstream to its beginning at the junction of the Tygart and West Fork Rivers in Fairmont, Wv. The methods described above were used. NGS also ran levels at various times along the entire river with the exception of a 27 mile stretch from Fredericktown to Point Marion, Pa. The NGS line left the river at California, Pa and followed a railroad to Uniontown and thence back to the river at Point Marion, near the Mason Dixon line. As mentioned earlier, the U.S. Army Corps of Engineers ran this section, but no field data or adjustment data can be found. As will be seen, there are problems with this line. The NGS ran their Basic Net A releveling line along generally the same route as before, but did not use the railroad. This survey, performed in 1983, followed highways and occasionally tied existing marks when the line came near them. This left long gaps along the river where the 1983 survey did not tie.

There were a number of river crossings which were required, and we did not have the specialized equipment required. I developed a special procedure to maintain the required accuracy. In several areas, we were able to cross at dams which had overhead structures. The procedure we used was as follows: Four tripods were set up, two on each side about 30 feet or less from bench marks. On each side, one tripod is occupied with a 1" theodolite, and a Wild precise target is placed on the other. First, the elevation is transferred from the bench mark to the trunnion axis of the theodolite by measuring direct and indirect zenith angles to four divisions on the rod (usually 1, 3, 5, and 7 foot marks). This is done because the rods are graduated every 0.02 feet, and interpolation would not be as accurate. It is estimated that this overly redundant method gives an accuracy of 0.002 feet. As a check, the rod is read at 90° and 270°. Then, simultaneous readings are taken on each side to the target on the opposite shore. Three or four sets of zenith angles are taken. After this, the theodolites are switched with the targets, and the procedure is repeated. Finally, EDM distances are measured one way for each set of tripods. This method is known as "Double Simultaneous Reciprocal Trigonometric River Crossing". Our first attempt at this procedure was done at Maxwell Lock and Dam, where we could also level across the piers of the dam as a check. The agreement between the two trigonometric readings was 0.012 feet over a distance of 760 feet. The mean of the two differed from the differential crossing by 0.003 feet (0.001 m). This compares with second order class I accuracy of 0.0029 m over that distance. This was the only direct comparison we had, but several loops were checked with this method. One loop, with a circumference of 6.4 miles, closed +0.009 feet (0.063 feet allowable). In another instance, a loop with a circumference of 18.5 miles had a closure of 0.007 feet (0.107 allowable). This loop was later split with another river crossing, resulting in two loops-one with a circumference of 11.4 miles and a closure of +0.004 feet, and the other with a circumference of 7.1 miles and a closure of +0.003 feet. We feel confident the method maintains the required accuracy for distances up to 1500 feet.

The first five bench marks in the run were existing USC&GS bench marks, and there were four at the southern end of the line. At the northern end, one of the five differed by 0.158 feet from its published elevation. Another differed by 0.028 feet from its published value. The other three checked within specifications. At the south end, all four checked well. This satisfied the requirement for at least two verified marks at either end. All leveling was single run using the double simultaneous procedure.

The first 26 miles ran along the left bank of the river, with a spur line river crossing at mile 13.2 to tie to a first order line on the right bank. Other than marks which had moved, agreement with both published elevations and 1983 observed heights was excellent. At mile 26, a river crossing was made at Lock 3. From there, the line traveled the right bank until mile 51, where the river was again crossed near California, Pa. There was a spur at mile 32 across a bridge to a bench mark on the left bank at Monongahela, Pa. This mark did not check, so a line was run from it back down river to where the river crossing was at mile 26, and another river crossing was made in the middle at mile 28 as a check. This loop is described above where I explained the river crossing procedure. Our results were proven. NGS was contacted and our findings were presented. I should mention that the bench mark in question agreed going upstream, but not going downstream. It appeared to be a jump in the line. NGS sent a level crew to check the mark. Their observed difference differed from ours by 0.0012 feet over the 1 mile distance between marks. It is believed that a spacer was used by NGS when coming into the mark, but not when going out. Our line continued along the left bank from California and crossed again at mile 61 at Maxwell Lock.This was the beginning of the section where NGS did not run. We ran along the right bank for 25 miles, crossed a railroad bridge to the left bank again, and ran another 4 miles to Point Marion Pa. At Point Marion, we ran a spur line across Lock 8 to a bench. This bench did not check well. We then checked into an adjacent bench. The results showed that the first bench had moved, since we agreed well with the 1983 determination. However, the discrepancy over the 37.7 miles was 0.195 feet versus published values, and 0.310 feet versus the NGS 1983 run. We reran 25.7 miles of the section (the remainder was proven with loops), and our rerun agreed with our first run within 0.116 feet. The allowable misclosure for 25.7 miles is 0.127 feet. This problem was discussed with NGS, but due to budget restaints, they are unable to field check this problem. As mentioned above, the U. S. Army Corps of Engineers had run this stretch of river supposedly to second order standards. However, the comparisons jumped around, and no conclusion was reached. From Pt. Marion to Fairmont, there were no further problems. Table 1 is a comparison of various adjustments of the level network. The input standard deviation was computed as 0.01 feet times the square root of the section length in miles. This corresponds to 2.4 mm times the square root of the distance in kilometers. This is rather optimistic, as reflected in the variances. However, no new values were computed, since it is only important that they are consistent between sections. The table shows the distance from the beginning as well as the published NGVD 1929 elevations. The first adjustment held only station N-343 fixed. The misclosure at station K-54, 132 miles away, was 0.356 feet versus the allowable 0.287 feet for that distance. However, no orthometric correction was made. Using the first run of the section which was rerun, the miscloure was -0.300 feet, and the orthometric correction is +0.072 feet, for a misclosure of -0.228 feet, within specs. The misclosure between the normal orthometric heights observed NGS in 1983 and our normal orthometric heights from B-417 to M-27 was 0.445 feet. This comparison used normal heights to account for the different paths taken. An F beside an elevation in the adjusted elevation columns denotes an elevation held fixed for that adjustment. The second adjustment held two stations fixed, one in the north, and one in the south. The third adjustment held 18 stations fixed. The last two adjustments dealt with the 1983 Basic Net A releveling. The first held the furthest north station from that survey fixed, while the last adjustment held one at each end.

One result of this leveling was that at least one station of each station pair had a known elevation. Using these elevations, the entire network was adjusted to compute orthometric heights for all the GPS points. Table 2 shows the results of four different adjustments. Adjustment 1 held one station near the center of the project. Adjustment 2 held three stations vertically, one at each end and one in the middle. The third adjustment held one station near the middle, and used geoidal heights from GEOID4. This held station was not the same as the previous one. We held a station in the area where we had good agreement between bench marks. Lastly, three stations were held and the geoidal heights were once again input. Remembering that this network covers 184 km north to south, the results are impressive. The addition of the geoid heights improves the results significantly. The most interesting conclusion is that, holding only one station fixed, it is possible to achieve an accuracy of 0.06 m over this large distance.

REGIONAL NETWORK
As different projects were being performed in the Pittsburgh area, an effort was made to provide continuity between the networks by connecting nearby stations. This resulted in a combined network stretching from Butler, Pa south to Clarksburg, Wv, and from Greensburg, Pa west to the state line. This network contains over 300 stations. The northern half of this network has been analized and adjusted. This consists of 231 stations. A free adjustment was done, hold-

ing one station fixed in latitude and longitude, and one station fixed in elevation. There were 520 baselines included in this adjustment. The estimated variance factor was 0.96, and there were no flagged residuals. The coordinates from this adjustment were used as input to GEOID4, a geoidal height estimation program from NGS. The NAD 1983 coordinates of 10 first order NGRS stations were then held, along with the same point held vertically. The estimated variance factor was 1.31, with a scale bias of 3.46 ppm. These are fairly good statistics given the size of the network and the number of points constrained. Lastly, 7 well distributed points with known elevation were constrained vertically, along with the same 10 points held horizontally. This had an estimated variance factoe of 1.21, a scale bias of 3.47 ppm, an azimuth bias of -0.29", and residual deflections of 0.05" in the prime vertical and -0.15" in the meridian. Table 3 lists the comparison between these three solutions at various benchmarks throughout the region. When one considers that the network extends 81 km east-west and 75 km north-south, the results are phenomenal. In the adjustment holding seven stations, the largest discrepancy is at 0681, 0.062 m. This was recently tied to a single bench mark, and the bench is not proven. This shows that GPS, with proper procedures and precautions, can be used to establish vertical control over relatively large areas sufficiently accurate for 2 foot contour interval mapping, and certainly for 5 foot contours.

SUMMARY

The need for careful preanalysis when planning control densification is seen. Most importantly, the consistency of control used is important. GPS determined elevations can be of an accuracy suitable for many projects if proper methods are used. Many agencies do not permit the use of GPS for vertical control. I feel this is due to past practices. One must realize that bench marks are subject to movement over time. The use of a geoidal height estimation program in the adjustment of the network is crucial to aiding in the estimation of orthometric heights. The NGS has a new model available, GEOID90, but it was not released in time for this study.

REFERENCES

Hayford, John F., 1904, _Precise Leveling in the United States 1900-1903 With a Readjustment of the Level Net and Resulting Elevations_. Government Printing Office

Hayford, John F., 1900, _Precise Loveling in the United States_. Government Printing Office

Olmstead, Frederick Law, 1911, _Pittsburgh-Main Thoroughfares and the Downtown District, Improvements Necessary to Meet the City's Present and Future Needs, A Report._, Pittsburgh Civic Commission

Arthur, U.N and Randall, R.H., 1925, _The Geodetic and Topographic Survey of Pittsburgh and Allegheny County_. Engineers Society of Western Pennsylvania

TABLE 1
RESULTS OF LEVELING ADJUSTMENTS

STATION	DIST.	ADJ1	ADJ2	ADJ3	ADJ4	ADJ5
N-343	0.00	+0.000F	+0.000F	+0.000F		
M-343	1.51	-0.009	-0.009	-0.005		
RVBRIDGE	4.52	+0.040	+0.038	+0.048		
K-343	3.59	-0.009	-0.010	+0.000F		
J-343	2.99	-0.005	-0.007	+0.000F		
W-133	16.64	+0.007	+0.004	+0.000F		
B-417	18.62				+0.000F	+0.000F
U-133	19.97	-0.001	-0.006	+0.000F		
DPW3035	22.29	+0.005	-0.000	+0.000F		
S-133RES	23.58	+0.027	+0.020	+0.025	-0.002	-0.005
R-133	25.35	+0.010	+0.001	+0.008	+0.000	-0.005
D-419	28.57				-0.008	-0.016
P-133	30.48	-0.001	-0.013	+0.000F	-0.010	-0.020
N-133	33.26	-0.010	-0.024	+0.000F	-0.005	-0.018
K-133	36.06	+0.030	+0.014	+0.037		
C-419	35.98				-0.005	-0.020
L-133	37.14	+0.030	+0.014	+0.037		
B-419	38.91				-0.009	-0.026
H-133	41.71	-0.004	-0.023	+0.000F	-0.005	-0.025
A-419	45.41				-0.007	-0.028
Z-417	48.07				-0.009	-0.030
Y-417	50.74				-0.001	-0.024
B-133	53.81	+0.011	-0.012	-0.007	+0.007	-0.017
A-133	55.52	-0.064	-0.088	-0.086	+0.044	+0.019
E-284	74.51	+0.039	+0.003	+0.013	+0.048	+0.008
H-132	88.33	+0.030	-0.014	+0.000F	+0.044	-0.007
F-120	97.72	+0.046	-0.005	+0.016		
D-120	108.88	+0.027	-0.027	+0.000F		
C-120	111.26	+0.042	-0.012	+0.015		
J-120	148.97	+0.121	+0.000	+0.000F	+0.120	+0.041
H-120	151.08	+0.036	-0.030	-0.027	+0.119	+0.039
STA. E	166.85	+0.057	-0.020	+0.000F	+0.134	+0.041
U-27	191.67	+0.084	-0.009	+0.000F		
M-27	210.03	+0.107	+0.000	+0.000F	+0.130	-0.001
L-27	210.19	+0.108	+0.002	+0.000F	+0.131	+0.000F
K-33	212.60	+0.106	-0.002	+0.000F		
K-54	212.78	+0.109	+0.000F	+0.000F		
Variance factor		2.02	1.74	2.74	1.74	2.20

Note: All values are in meters, distances in km. F denotes fixed station.

TABLE 2
COMPARISON OF HEIGHTS DETERMINED BY DIFFERENT ADJUSTMENTS FOR MON RIVER GPS NETWORK

GPS#	1-LEV	2-LEV	3-LEV	4-LEV
0537	-0.906	-0.044	-0.023	-0.044
0539	-0.753	HELD	+0.006	HELD
0540	-0.690	+0.004	-0.034	-0.022
0541	-0.679	+0.006	-0.035	-0.020
0542	-0.646	+0.003	-0.019	-0.008
0545	-0.607	+0.000	-0.006	+0.002
0547	-0.556	-0.004	+0.001	+0.006
0549	-0.589	-0.047	+0.001	-0.009
0550	-0.426	+0.020	+0.002	+0.018
0552	-0.386	+0.020	+0.025	+0.034
0555	-0.273	+0.027	-0.015	+0.010
0556	-0.309	-0.001	HELD	+0.013
0558	-0.212	+0.037	+0.027	+0.044
0559	-0.192	+0.022	+0.021	+0.035
0583	-0.252	-0.006	+0.030	+0.033
0588	-0.269	-0.014	+0.047	+0.038
0589	-0.174	-0.024	-0.011	-0.001
0590	-0.156	-0.017	-0.005	+0.005
0592	-0.048	+0.006	-0.003	+0.012
0422	HELD	HELD	-0.016	HELD
0593	+0.005	-0.005	-0.022	-0.006
0426	+0.090	+0.044	+0.009	+0.023
0595	+0.104	+0.034	+0.007	+0.026
0596	+0.200	+0.062	+0.007	+0.027
0597	+0.212	+0.073	+0.015	+0.035
0599	+0.133	+0.027	-0.051	-0.044
0600	+0.310	+0.080	+0.017	+0.029
0615	+0.312	+0.053	+0.019	+0.021
0603	+0.332	+0.067	+0.020	+0.021
0604	+0.291	+0.036	+0.012	+0.007
0605	+0.277	+0.018	-0.003	-0.010
0607	+0.251	HELD	+0.026	HELD
0608	+0.299	0.000	+0.018	-0.010

All values are in meters.

TABLE 3
COMPARISON OF HEIGHTS DETERMINED BY DIFFERENT ADJUSTMENTS FOR REGIONAL GPS NETWORK

GPS#	*	1-LEV	2-LEV	3-LEV
0053	B	0.381	0.010	0.023
0055	C	0.147	0.020	0.041
0057	C	-0.115	-0.083	HELD
0225	A	-0.003	-0.025	-0.027
0226	A	-0.040	-0.032	-0.035
0231	A	-0.103	-0.022	-0.021
0282	B	0.199	0.043	0.045
0340	A	-0.154	-0.032	-0.026
A437	A	-0.277	-0.012	HELD
0537	A	-0.145	-0.027	-0.023
0539	A	HELD	HELD	HELD
0540	A	0.086	-0.012	-0.013
0541	A	0.074	-0.036	-0.037
0543	A	0.109	-0.016	-0.015
0545	A	0.141	-0.005	-0.001
0547	A	0.181	-0.012	-0.006
0549	A	0.148	-0.005	0.001
0550	A	0.315	0.013	0.000
0552	A	0.352	0.029	0.020
0555	A	0.476	0.004	HELD
0556	A	0.436	0.017	0.016
0628	A	-0.176	-0.009	-0.005
0631	A	-0.186	-0.008	-0.003
0635	A	-0.145	0.021	0.028
0640	A	-0.212	0.003	0.012
0642	A	-0.150	0.015	0.023
0657	B	0.000	0.054	0.053
0665	A	-0.132	0.004	0.007
0681	C	0.119	-0.062	-0.062
0776	B	-0.048	0.026	0.027
0777	A	0.013	-0.017	-0.018
0816	A	0.624	-0.030	HELD
0817	A	0.654	-0.088	HELD
0818	A	-0.003	0.016	0.008
0819	A	0.012	0.022	0.012
0820	A	0.009	0.026	0.016
0822	A	0.009	0.029	0.019
0823	A	0.106	0.022	HELD
0824	A	0.386	-0.067	-0.039
0833	A	-0.150	0.027	0.035
0834	A	-0.153	-0.010	0.003

*QUALITY CODE: A=high quality bench mark on stable structure or recently leveled.
B=good quality bench mark, not checked
C=questionable quality, bm in conc post, not checked.
All values are in meters.

RESULTS OF WISCONSIN HIGH
PRECISION GEODETIC NETWORK SURVEY

Paul J. Hartzheim
Wisconsin Department of Transportation
Madison, Wisconsin 53707

Larry D. Hothem
National Geodetic Survey
Space and Physical Geodesy Branch
Rockville, Maryland 20852

Dale W. Kyle
Western Geophysical
Houston, Texas 77042-4299

ABSTRACT

A high accuracy network using GPS differential survey techniques was established in Wisconsin. This network, called the Wisconsin High Precision Geodetic Network (WHPGN), consists of 80 primary stations located 50 km apart throughout the state and 18 secondary stations which form ties to the NGRS.

The GPS survey was performed in two phases. In Phase 1, GPS observations were observed on 5 WHPGN stations simultaneously with stations of the CIGNET. Western Geophysical used their proprietary orbit adjustment software, SONAP, to process data between the CIGNET and 5 WHPGN stations to Order A accuracy standards of 1 part in 10,000,000 on a daily basis and in a 3-day arc solution. In addition, Western's proprietary software, AIMS, used the precise ephemerides to process Phase 1 data. Phase 2 GPS observations were observed on all stations and reduced using precise ephemerides. Coordinates were selected from a Phase 1 adjustment and held fixed in the adjustment of Phase 2 data to meet Order B accuracy standards of 1 part in 1,000,000.

This paper will discuss the WHPGN, formation of contract specifications, GPS observational data reductions, analysis of adjustments, relationship of coordinate systems to this project, comparisons between NAD 83 published and WHPGN coordinates, and field checks.

A NEW PLANE TABLE SYSTEM---- CG-PLANE TABLE

Susumu Hattori
Dep. of of Eng., Fuc.of Eng., Fukuyama Univ.
985 Aza-Sanzo, Higashimuracho, Fukuyama City,
Hiroshima Pref., 709-02, Japan

Hiroyuki Hasegawa and Kouhei Uesugi
Pasco Corporation
2-13-5 Higashiyama, Meguro-ku, Tokyo, 153, Japan

ABSTRACT

A CAD-based new plane table surveying system is developed for topographic surveying as a major subsystem of a digital mapping system. This system (named CG (computer graphics)-plane table) consists of a total station interfaced with a lap-top personal computer and software of a CAD system and other applications. With the system surveyors observe objects, edit figures and compile a map up to a near finish at a survey site without storing data in a data collector. According to experiences the system is estimated to highly improves both the reliability and efficiency of topographic surveying compared to conventional plane tables or total stations. This comes from the following advantages:
1) No ambiguities remain in a detail survey, because measurements are always monitored with a graphic display.
2) Either a control survey or a detail survey can come to the first depending on a situation or a time schedule, since all positional data are stored in a digital form and can be retrieved in any time for recompilation.
3) Data are easily connected to other data resources, because the common CAD supervises all the surveying works consistently.

INTRODUCTION

The need for digital topographic maps is rapidly increasing. They are applied to base maps for CAD systems in construction design, for geographical information systems or for car navigation systems. Digital maps for large areas are usually produced by the digital mapping technology in aerial surveying, while for small areas they are produced by conventional ground surveying. And large scale aerial surveying often requires supplementary ground surveying due to many occlusions included in photographs. Ground surveying is executed with a conventional plane table system or a total station system.
A conventional plane table is a simple insturment to sketch objects at survey sites together with a tape and an alidade. But with this instrument it is essencially impossible to expect high precision. Further in compiling digital maps raster-vector convertion from analogue drawings to digital ones is necessary in the office work, which is an undisregardable burden and takes much time.
The development of total stations has resolved some of

the difficulties inherent in a conventional plane table system (Kennie 1990). The total station is a system linking a EDM (Electromagnetic Distance Meter) to a digital theodlite and usually equipped with a data collector as an output device. The system can yield 3-D coordinates of objects in a digital form and store them in a data collector. These data are brought back to the office to be compiled with a CAD to a finish map (Froelich 1988, Turner 1987).

However as a matter of fact the efficiency and reliability of a topographic survey with a total station is not much more improved than with a conventional plane table, because
1) Many intermediate files must be made until finish products are obtained. Transportation and confirmation of data from a data collector to a computer and interactive compilation of a map takes much time. Especially this is the case when compiling maps with aerial survey data and ground survey data which must be merged, because they have usually diferent formats.
2) Umbiguous or erroneous measurements may be found at the office, since a total station has no monitoring display for visual confirmation, which may require remeasurement at the surveying site afterwards.
and
3) In urban areas 3-D maps are often required for the data base describing each floor use in tall buildings from underground to the top. But data collectors lack the data management function enough for 3-D maps, i.e., detailed description of land use and swift data transportation to an data base.

In order to solve the above drawbacks we have recently developed a CAD-based new plane table surveying system (we call the system CG-plane table surveying system or more concisely CG-plane table). It consists of a total station linked to a personal lap-top computer and the software of a CAD and necessary application programs. With the system surveyors can measure object points coordinates, edit lines and figures and compile a map to the near finish under control of the CAD consistently at the survey site. In this system, a total station can be regarded not as an independent instrument but rather a 3-D input device for the CAD just like a usual digitizing table.

The system is featured by functions:
1) On-line adjustment in control surveys
2) On-line drawing and editing of lines and symbols
3) In-site compilation of maps to near finishes
4) Easy data linkage to other data resources and data bases

The system is already now in successful use in Pasco Co.. In the following the constitution and functions of the CG-plane table are discussed and the working efficiency is compared with a conventional plane table and an existing total station system.

CONSTITUTION OF CG-PLANE TABLE SURVEYING SYSTEM

Fig.1 shows an outlook of the CG-plane table surveying system. Since most of total stations are provided with serial interfaces, it was of no need to redesign existing

(a) Total station (TOPCON GT3) and personal computer (TOSHIBA J3100)

(b) Reflecting mirror

Fig.1 Outlook of CG-Plane Table Surveying System

total stations to meet our porpose. What we have been mainly devoted to is developing application software as well as an interface driver.

Hardware

Total station TOPCON GT3 and ET-2 are alternatively in use at the moment. Any total stations are usable as long as they have an RS232C interface.

CAD system AUTODESK AutoCad GX-III, Release 10 is adopted. This has ample 3-D processing commands and is already very popular in a personal CAD market.

<u>Personal Lap-top computer</u>   TOSHIBA J-3100 and COMPAQ SLT are alternatively in use from the viewpoint of sufficiently long driving duration ability by batteries and fast data processing performance. A co-processor and a more than 20 Mbytes hard disk are necessary to run AutoCad.
<u>Reflecting mirror</u>   As a hardware instrument a special-purpose reflecting mirror is developed. Existing prism mirrors are too large and less portable to set on many object points consecutively. As shown in Fig.1(b) the reflecting mirror is icosahedron made of plastics on which reflecting metal is stuck. One side is 15 mm long (i.e., the inscribed globe being 23mm across in diameter). The measuring distance limit is about 50 m. Distance errors in measurements due to the thickness of the mirror is corrected with a program.

<u>Software</u>
Some of the existing software are transplanted into the system and others are newly developed. The former include usual adjustment programs for control surveying and fonts and symbols in use for map compilation for the CAD. The software developed for the system are as follows:

<u>Interface driver program</u>   This is to interface between the personal computer and the total station. The program was made using a device driver package, ADI (Autodesk Device Interface) which is offered by Autodesk Co..

<u>Intialization program for the total station</u>   This is to wake up the total station and to set initial values.

<u>Resection program</u>   This is to determine instrument positions. With the CG-plane table we have two ways to fix instrument posisions. One is setting the instrument on the predetermined points by usual control surveys and orienting it to the reference directions. And the other is determining a position and a reference azimus of the instrument by resection which is placed at an arbitrary position.
The resection method in conventional plane table surveying is known as the three points problem. Some computational solution programs for the resection problem were published for hand-held computers (Elhassan 1986, Maier 1982). But our program is far sofisticated because of higher ralculation performance of the lap-top compouter in that:
*it calculates approximations of unknowns automatically. Usually they used to be input by an operator.
*it conducts simultaneous adjustment of slope distance and horizontal and vertical angles by the bundle adjustment which permits errors to control points coordinates as well as observations.
The program needs complicated computations, therefore is unexecutable in handheld computers.

<u>Intersection program</u>   (in tentative use)   Most of object point identification in detail surveys are executed by the method of radiation. But for objects unaccessible to mirror-holders the method of intersection becomes necessary.

TOPOGRAPHIC SURVEYING PROCEDURE WITH CG-PLANE TABLE

Surveyors can proceeds with topographic survey more efficenly with the CG-plane table system than by the usual way owing to the CAD function and newly developed software. The following is the surveying procedure Pasco Co. currently adopts.

Control survey

Control surveys are executed in usual ways, i.e., by traversing, triangulation and/or trilatelation. Observations can be checked and adjusted immediately in survey sites.

Detail survey

A total station is set on control points the coordinates of which are known, or set on an arbitrary point the coordinates of which are determined by resection. Most of object coordinates may be measured by the method of radiation, and some by the method of intersection. Surveyors display observed points on a screen and draw lines and symbols with the CAD.

Drawing files produced by aerial surveys, if any, are easily handed over or merged to drawing files in the ground survey in a common format, which facilitates the following works.

There are two surveying modes in selection of the coordinate sytem. One is the surveying with each local origin being at the instrument position where the total station is set. Data files are brought back to the office and compiled to the finish in the global coordinate system. This procedure is possible because instrument position coordinates are preserved in a digital form and can be retrieved any time. The other is the surveying with a unified global origin, as has been conventionally executed.

Actually it is ideal from the precision point of view to execute the control survey after the detail survey. The control survey results together with resection data should be adjusted simultaneously in a global network. Map compilation should follow that. This procedure is now under planning.

The map data are , if necessary, transfered to a plotter in the headquarter through a telephone line for a quick drafting.

DIGITAL MAPPING SYSTEM IN PASCO CO.

Pasco Co. has now integrated the CG-plane table surveying system to the Pasco digital mapping system (MAPCAD) (Hasegawa 1990), which consists, as shown schematically in Fig.2, of subsystems:
 CG-plane table (MAPCAD 10),
 Map compilation sytem with digitizers(MAPCAD 20),
 Photogrammetry (MAPCAD 30), Drafting system (MAPCAD 40) and
 GIS, Data base and Computer vision (MAPCAD 50).

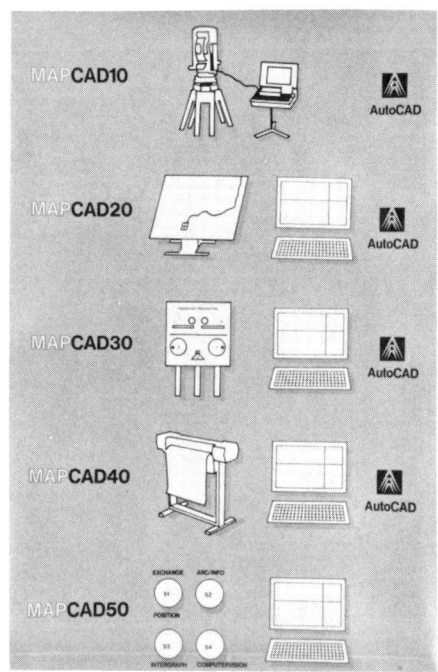

Fig.2 Subsystems of PASCO Digital Mapping System, MAPCAD
MAPCAD 20 to 40 are controled by the common CAD (AutoCAD).

EFFICIENCY COMPARISON OF CG-PLANE TABLE SURVEYING WITH CONVENTIONAL PLANE TABLE SURVEYING

The authors have experienced twice the map production projects in Kasumigaseki, gavernmental office block in Tokyo, in 1971 and in November 1990. The area is 30,000 m2 and the map scales are both 1:500. The former was to make an analogue urban land use map, while the latter was to make a data base for an urban GIS (geographical information system). Fig.3 shows a perspecitve expression of the area produced in the latter project.

Fig.4 lists up the works included in two projects. Main portion of the area was plotted with aerial surveys. The old project was done in a conventional way with an analogue plotter, while the new project was executed by the digital mapping method with an analytical plotter. Both required supplementary surveys. The old project was done by conventional plane table surveying, and the new one was done with the CG-plane table. From these experiences we can estimate the expected time consumed in the supplementary survey if the conventional plane table is employed in the new project. We now neglect the effect due to the technological progress in aerial surveying in 20 years.

The supplementary ground survey with the CG-plane table in the new project was executed at the rate of 10,000 m2/day/unit. In Pasco Co. a serial survey works are supposed to be executed in a unit of three to four persons.

In contrast to this, in the old project the ground survey was done with the conventional plane table at the rate of 7,000 m2/day/unit. But if the conventional plane table

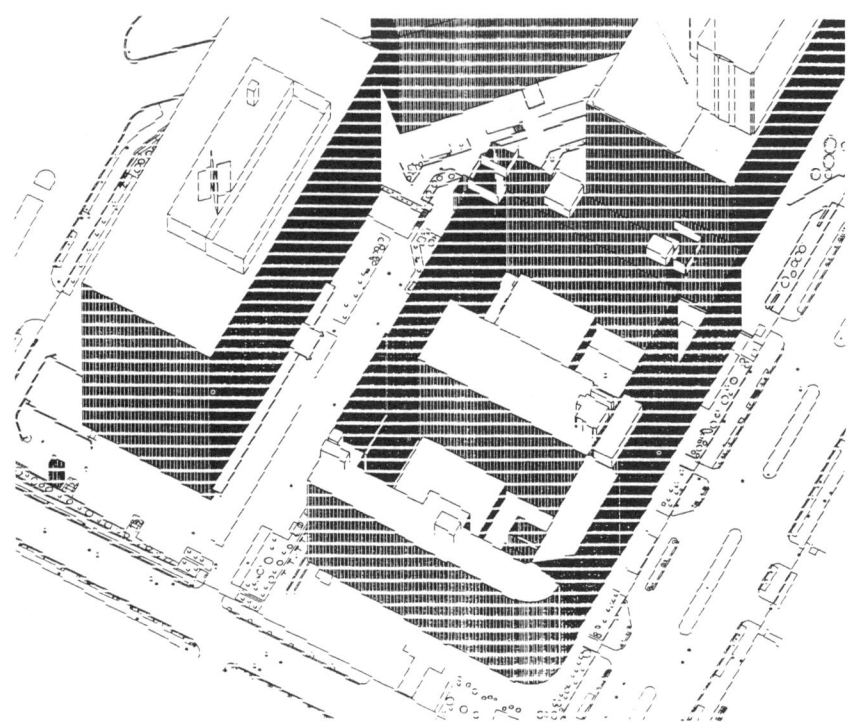

Fig.3 Perspective view of the area produced by the Kasumigaseki project in 1990

PROJECT in 1971	PROJECT in 1990

PHOTOGRAMMETRY
   Control Survey
     (Theodlite + Level)
   Placing Aerial Sygnals
 	
   Photographing
   Plotting
     (Analogue Plotter)
	Control Survey
	   (CG-plane table)
	Placing Aerial Signals
	   (20% reduced in number to the left.)
	Photographing
	Digital Mapping
	   (MAPCAD 30 with an Analytical
	                                Plotter)

SUPPLEMENTARY GROUND SURVEY (Conventional Plane table)	(CG-plane table)

COMPILING & DRAWING MAPS
   (Conventional Scribing)
	(MAPCAD 20 with a map compilation
	 system; MAPCAD 40 with a drawing
	 table)
	
	DATA BASE CONSTRUCTION
	   (MAPCAD 50 with a data base
	    management system)

Fig.4 Outline of Kasumigaseki projects

surveying replaces the CG-plane table surveying, map digitization is necessary to prepare data for the succeeding works(compilation of finish maps and production of the data base), the rate of which would be 7,000 m2/day/unit. Thus the actual rate would be 3,500 m2/day/unit. This means the CG-plane table surveying system is about three times more efficient than the conventional plane table.

CONCLUSION

The newly developed CAD-based plane table surveying system is discussed. The system called CG-plane table consists of a total station linked to a lap-top personal computer. Surveyors observe 3-D coordinates of objects and draw lines and symbols on a computer screen at survey sites with the CAD. Since maps can be compiled up to near finishes at the site without bringing data back to the office, the survey efficiency and reliablity are much improved. The efficiency with the CG-plane table is estimated to be three times higher than with the conventional plane table according to actual survey projects.

REFERENCES

Elhassan,I.M.1986:An Analytical Solution of the Resection Problem. Journal of Surveying Engineering,Vol.112, No.1

Froelich,H. and Wolfgang Jez 1988: Halt-Praxisorientierte PC-Software zur Auswertung terrestrischer Lagevermessungen, Zeitschrift fuer Vermessungwesen, Heft2,pp.72-79

Hasegawa,H.,J.Rogers and A.Follett 1990: MAPCAD - An AutoCAD Based and ARC/INFO Oriented Digital Mapping System, ISPRS Symposium, Tsukuba

Kennie,T.J.M. 1990:Electronic Angle and Distance Measurement, a topic in Engineering Survey Technology edited by T.M.J. Kennie and G. Petrie, Blackie and Son Ltd.,pp.7-47

Maier,U.1982:Ein Taschenrechnerprogramm zur Punktein- shaltung und freien Stationierung, Algemeine Vermessungs Nachrichten, Heft4,pp.158-167

Turner, H.1987: A Model to Integrate Data Collectors, Surveying and Mapping, Vol.47. No.2, pp117-123

ELEVATION DIFFERENCE RECOVERY TESTS FROM GPS OBSERVATION AND GRAVIMETRY

K. Jeyapalan and E. S. Erck
Iowa State University
Ames, IA 50011

ABSTRACT

GPS gives ellipsoidal heights, where as in engineering applications, orthometric heights above a geoid or mean sea level are required. Transformation of ellipsoidal heights to orthometric heights requires precise geoidal undulation. The spherical harmonic expansion of global gravity anomalies can be used to give the major portion of the geoidal undulation, which is called the global undulation. Utilizing the local gravity observations, local gravity anomalies can be computed by subtracting global gravity anomalies. These local gravity anomalies can then be used to predict the local component of the undulation. For three stations, in central Iowa, separated by distance of 15 Km to 130 Km the improvement in orthometric height by local undulation was found to be about 0.5 m.

INTRODUCTION

The development of the Global Positioning System (GPS) offers more accurate and efficient positioning potential than that offered by conventional methods. GPS procedures are suitable, for example, in navigation where a high degree of accuracy is required mainly in obtaining horizontal coordinates. However, in engineering and some other fields of study, vertical coordinate accuracy remains a priority. These orthometric height coordinates of elevations are unlike the horizontal since they are based on a dynamic or natural systems having physical rather than an analytical datum. A problem arises in the transformation of the analytical GPS geodetic heights to natural heights above sea level (approximately the geoid) required for engineering applications.

Geoid modelling can provide the solution for this height transformation problem. Recent research in geoid modelling has generally utilized (non-GPS) satellite altemetry data and surface gravity data. Global geo-potential models have been developed through spherical harmonic expansions. Geoidal undulations from these models have often been further delineated using surface gravity data near the station of interest. The models of Fury and Kearsley improve global models expanded to degree and order 22 and 180, respectively, by incorporating local gravity data in 20 square degree and 1 degree radius areas, respectively. The model developed by Rapp depends on a global geo-potential model truncated to a high degree and order of 360. The model of Jeyapalan and Erck, described in this paper, adds global geoidal undulation components from a high degree and order 360 geo-potential model and a small local geoid undulation from 0.5 radius spherical cap. The objective of

this paper is to give the orthometric height or elevation differences recovered at 3 central Iowa stations (separated by a distance of 15 Km to 120 Km) by a stand alone GPS receiver (Navcore) and differential GPS receivers (Ashtech) utilizing these recently developed geoid modelling resources. The research will be presented under the following headings:
- o Modes of operation of GPS
- o Theory of gravity geoid model both global and local
- o GPS data acquisitions and processing
- o Gravity data processing
- o Results and conclusions

1.0. MODES OF OPERATION

GPS positioning is performed in one of three modes, depending on the use and accuracy required. The most basic is the stand alone mode (7). The GPS receiver software resects the geocentric Cartesian and geode xc position of its single antenna with respect to the visible satellite constellation. The stand alone mode is mostly used for navigation purposes. The differential or relative mode utilizes phase differences of the satellites' emitted radio signals to give the differences in geocentric Cartesian coordinates between the two antennae. This mode achieves greater accuracy than the stand alone mode, but requires more data postprocessing. This level of accuracy has been extended into a kinematic mode (13). It differs from the differential mode by its greater speed in producing antennae separation components. The stand alone mode was implemented in this investigation by the Navcore receiver which was used.

2.0. THEORY OF GRAVITY GEOID MODEL BOTH GLOBAL AND LOCAL

For GPS to be useful in a natural coordinate system, the geodetic antenna position coordinates produced by the receiver must undergo geodetic to natural coordinate transformation (8). Since GPS heighting is the focus of this study, the height transformation equation, $H = h - N$) is examined, where H is the orthometric height, h is the ellipsoidal height and N is the geoidal undulation. With h determined as accurately as possible, an equally accurate solution for N is desired. Cravimetric methods for the geoid modeling or N determination are preferred over the less direct and more field time intensive astronomic methods. The gravimetric geoid modeling is accomplished by the residualization and superposition of global and local geopotential components (15). Similar methods exist for the modeling of the deflection quantities ξ and η, but their exclusive detail is beyond the heighting focus of this study.

2.1. Global Geopotential Components

The geoid must first be considered as a gravity equipotential surface for the entire Earth. It is therefore necessary to determine the entire Earth. It is therefore necessary to determine the entire Earth's disturbing or anomalous gravity potential T at the point on the ellipsoid with the same horizontal coordinates as the point of interest. This procedure requires much gravity (acceleration) data to fit a long spherical harmonic series. Once complete, the global geoidal undulation and gravity anomaly are derived.

The spherical harmonic expansion of the disturbing potential T is

$$T(\theta,\lambda) = \frac{GM}{R}\sum_{n=2}^{n_{max}}\sum_{m=0}^{n}(C_{nm}\cos m\lambda + S_{nm}\sin m\lambda)P_{nm}(\cos\theta)$$

(2.1)

from (15). G is the universal gravitational constant, M is the mass of the Earth, and R its mean radius. C_{nm} and S_{nm} are the fully normalized potential coefficients, corrected for the ellipsoid, and P_{nm} is the fully normalized Legendre function. These quantities are of degree n and order m. The infinite series is truncated to degree and order n_{max}. (See (3) and (11).) The lower degree and order harmonic coefficients are obtained by least squares fitting of (non-GPS) satellite altimetry data. The higher ones are similarly obtained by surface gravimetry all across the globe.

Derivation of the desired global quantities follows. The well known Brun's formula

$$T = N\gamma$$

(2.2)

and normal (theoretical) gravity at a spherical surface

$$\gamma = \frac{GM}{R^2}$$

(2.3)

equations are substituted in the previous series to give the global geoidal undulation

$$N(\theta,\lambda) = R\sum_{n=2}^{n_{max}}\sum_{m=0}^{n}(C_{nm}\cos m\lambda + S_{nm}\sin m\lambda)P_{nm}(\cos\theta)$$

(2.4)

A close derivation of the fundamental equation of physical geodesy

$$\Delta g = -\frac{\partial T}{\partial r} - \frac{2T}{r}$$

(2.5)

(where r is the radial Earth direction) forms the derivation of the global gravity anomaly expansion from Equation (2.1).

$$\Delta g = \frac{GM}{R^2}\sum_{n=2}^{n_{max}}\sum_{m=0}^{n}(n-1)(C_{nm}\cos m\lambda + S_{nm}\sin m\lambda)P_{nm}(\cos\theta)$$

(2.6)

See (15, pp. 155-157) for Equations (2.1) through (2.6). These series expansion equations can compute the longer wavelength components of the disturbing potential, geoidal undulation, and gravity anomaly. Higher degree and order expansions broaden the frequency spectra of these surfaces towards higher frequencies.

2.2. Local Geopotential Components

It is desired that the accuracy of the global geoidal undulation be improved by also analyzing local geopotential (high frequency) gravity anomalies and geoidal undulations in the area of interest. These free air gravity anomalies (5) come from observations rather than opened or closed form equations. Global gravity anomalies are computed by Equation (2.6) at the horizontal locations of the observations. The computed anomalies are then subtracted from the free air anomalies to produce residual gravity anomalies.

Stokes' formula integration follows, transforming residual gravity anomalies surrounding a station of interest to a local or high frequency geoidal undulation component. The formula is based on Stokes' theorem which states that the gravity potential, in the exterior space of an enclosing level surface of a rotating mass, is uniquely determined.

The general formula (6) is

$$N = \frac{R}{4\pi\gamma}\iint \Delta g S(\psi)\,d\sigma$$

(2.7)

where γ is the average normal gravity over the ellipsoid, Δg the gravity anomaly, and ψ is the spherical distance between the point of interest and a surrounding residual gravity data point. Stokes' function is

$$S(\psi) = \csc\frac{\psi}{2} - 6\sin\frac{\psi}{2} + 1 - 5\cos\psi - 3\cos\psi\ln(\sin\frac{\psi}{2} + \sin^2\frac{\psi}{2})$$

(2.8)

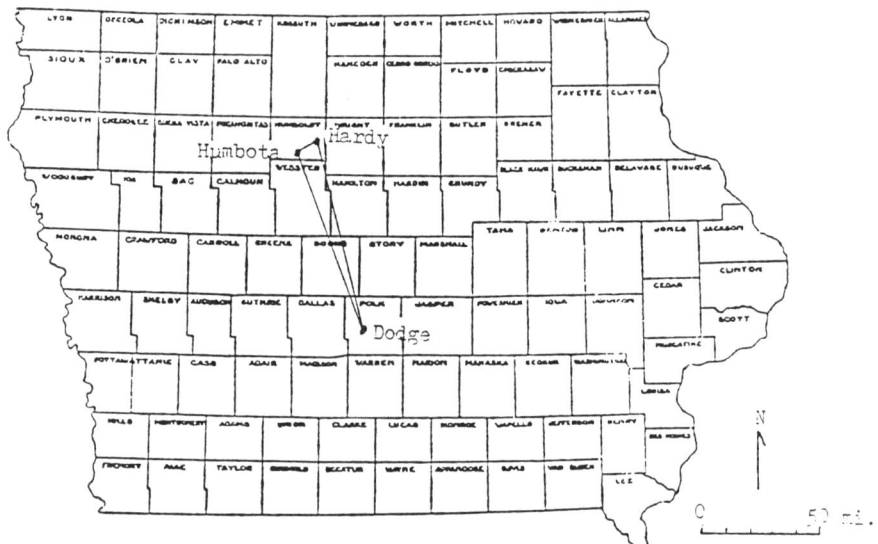

Figure 3.1. Three NGS total stations

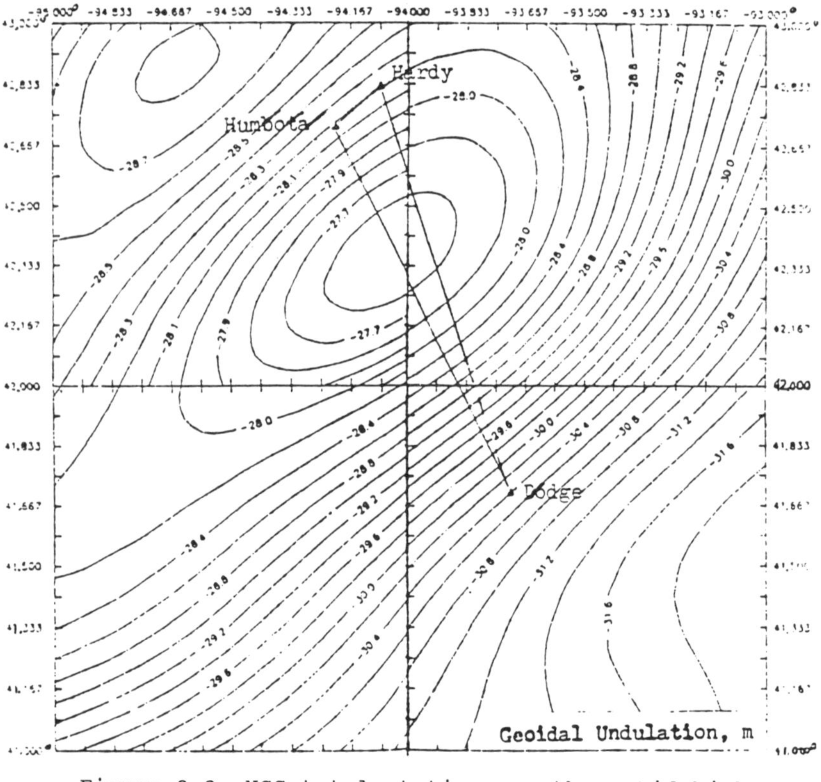

Figure 3.2. NGS total stations on the geoid high

and dσ is a surface element. When the enclosing surface is a spherical cap (around a point of interest i) having concentric radial compartments k, then the local geoidal undulation component is N_{Li}. From (6, pp. 118-119):

$$N_{Li} = \sum_k c_k \Delta^2 g_k$$

(2.9)

where

$$c_k = \frac{R(\alpha_{2k}-\alpha_{1k})}{4\pi\gamma} \int_{\psi_{1k}}^{\psi_{2k}} S(\psi)\sin\psi\, d\psi$$

(2.10)

Therefore

$$N_{Li} = \sum_k \Delta^2 g_k \frac{R(\alpha_{2k}-\alpha_{1k})}{4\pi\gamma} \int_{\psi_{1k}}^{\psi_{2k}} S(\psi)\sin\psi\, d\psi$$

(2.11)

The average residual gravity in a compartment is $\Delta^2 g_k$. The concentric radial compartments are bounded by radial limits ψ_{1k} and ψ_{2k} and azimuth limits α_{1k} and α_{2k}. N_{Li} is the correction that improves the accuracy of (by superposition to) the global geopotential geoidal undulation when the "enclosing" surface encloses a portion of the globe.

3.0. GPS DATA ACQUISITION

In late July and early August, 1988, Navcore GPS data were acquired at three stations in central Iowa for which National Geodetic Survey (NGS) total (three dimensional) control was established. Figure 3.1 shows they are arranged in a triangle such that one leg is much shorter than the others. The two longer legs also traverse a gradient of the geoid high that is associated with the Midcontinent Gravity High. See Figure 3.2.

The GPS data acquisition program specified four requirements regarding data collection times and intervals, maximum recording GDOP, and antenna placement. The data were collected at each station on three separate days during their 4 (approximate) hours of visibility. The recording interval was 15 seconds per geocentric Cartesian position sample. Recording was interrupted when the GDOP exceeded 10. The Navcore antenna was set directly on the point of observation in order to minimize reflected signal reception.

In Spring 1989, Ashtech GPS differential data were acquired at these stations and number of other stations in central Iowa. The observations were done in static mode and

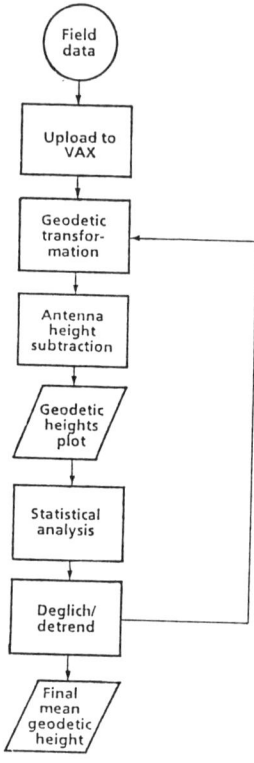

Figure 3.3. GPS data processing

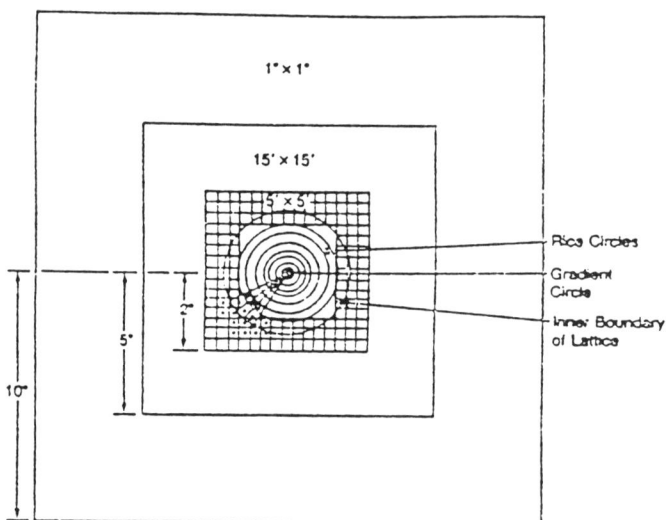

Figure 4.1. Spherical cap of 400 square degrees

the data were collected for about 2 hours at each station with a recording interval of 20 seconds.

3.1. GPS DATA PROCESSING

After decoding, the position data underwent geodetic transformation and statistical analysis on the ISU VAX 11/780 computer. Figure 3.3 summarizes these processes.

The floppy disks from the Navcore recording (Compaq) computer were uploaded to the VAX. The geocentric Cartesian position data were transformed to geodetic height above the GRS80 ellipsoid, or the NAD83 geodetic datum (1). The height of the antenna's electrical center above the point was then subtracted, prior to plotting all the geodetic heights.

Statistical analysis of the geodetic height data followed. For each station, means and standard deviations of the heights were taken. Heights that were not within two standard deviations of the mean were accepted in a final population only if they were within three standard deviations of the mean of the population without them.

The plots were examined for severe spikes or trends. Such data were rejected and the transformation and statistical process repeated.

The Ashtech data collected were adjusted using the three dimensional adjustment software "Geolab" with sufficient elevation and horizontal control points to give the ellipsoidal height. The station "Dodge" was fixed in the adjustment.

4.0. GRAVITY DATA PROCESSING

Gravimetric geoid modeling was accomplished by two methods.

The first method (3) utilizes a low degree and order global geopotential model and a large spherical cap. The GEM-10 geopotential model of degree and order 22 is greatly improved by a 400 square degree spherical cap (Figure 4.1). The process was done at NGS because of the large gravity data base required for the spherical cap (4).

The other method reverses this model and cap size. A high degree and order global geopotential model, the OSU86F, is improved by the superposition of a local geoidal undulation from a small spherical cap. Since the degree and order of the global model is 360, then the spherical cap radius is 180/360 of 0.5 degrees. (See (8).) This was done on the ISU VAX 11/780 and NAS AS/9160 mainframe computers (Figure 4.2).

The first step in this gravity data processing was the selection of the free air gravity data from the Iowa Geological Survey Bureau (IGSB) gravity data base (5) for the vicinities of the three stations. Since the free air gravity data were reduced to the GRS67 ellipsoid, a

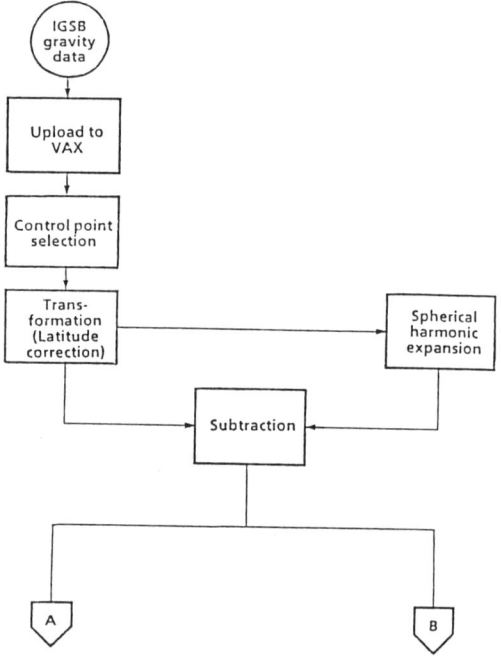

Figure 4.2. High expansion/small cap gravity data processing

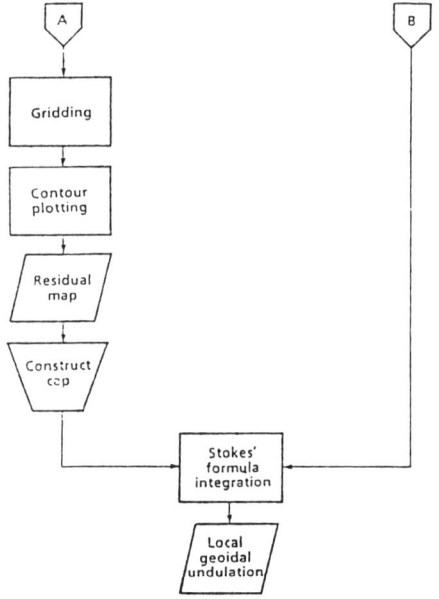

Figure 4.2. (continued)

correction had to be applied to both the latitude and gravity anomaly (14). The latitude correction first required a transformation of the latitude from the GRS67 ellipsoid to geocentric coordinates (2).

$$\phi = \tan^{-1}[(1-e_{167}^2)\tan\phi_{67}]$$

(4.1)

This equation was used in reverse with the GRS80 first eccentricity to obtain the GRS80 latitude (9).

$$\phi_{80} = \tan^{-1}[\tan\phi/(1-e_{180}^2)]$$

(4.2)

The gravity latitude correction was achieved by adding the following (in mGal) to the GRS67 gravity anomalies. (See (10).)

$$\Delta g = -.8316 - .0782\sin^2\phi_{67} + .0007\sin^4\phi_{67}$$

(4.3)

Therefore

$$\Delta g_{80} = \Delta g_{67} + \Delta g_{6780}$$

(4.4)

where Δg_{67} and Δg_{80} are the GRS67 and GRS80 gravity anomalies respectively. The GRS80 locations and values of the free air gravity data result.

These locations are the input to the spherical harmonic expansion computer program (12) that generates global geopotential gravity anomalies and geoidal undulations at the locations. The program cannot be reproduced in this writing. The global geopotential model is the OSU86F to a degree and order expansion of 360. It utilizes a 30 square minute grid of mean gravity anomalies from satellite altimetry and some gravity data from land, including the United States (11). The computer program was adapted to run on the NAS AS/9160 computer. Its output was subtracted from the free air gravity data to yield the residual gravity data.

Figure 4.2 indicates a bifurcation at this point in the processing flow. The left branch details the interpretive intervention required in the analysis of the residual gravity data. The data are gridded on the NAS AS/9160, using local quintic polynomials, prior to contour plotting. The spherical caps are constructed on these residual

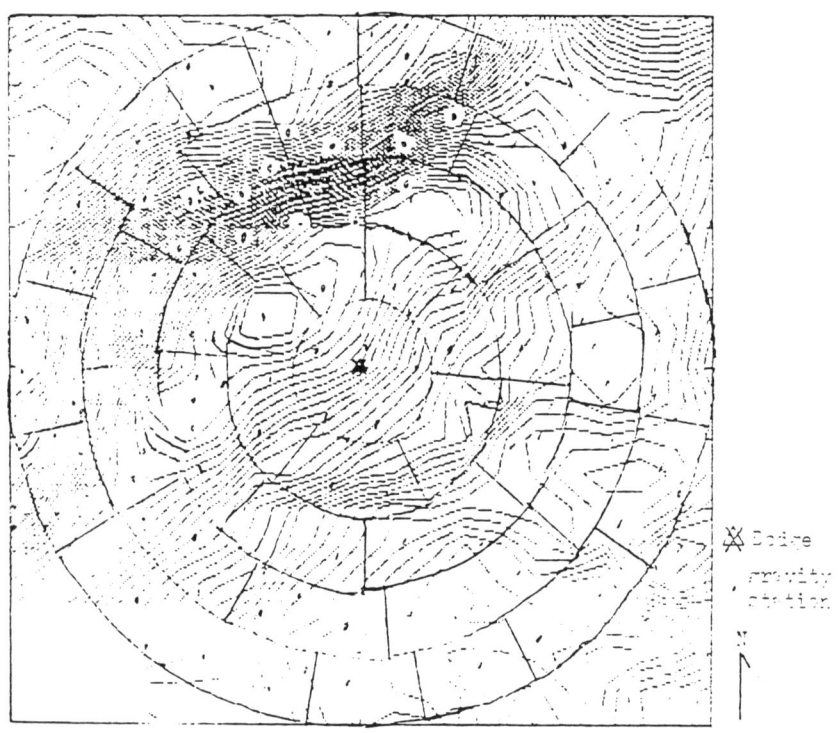

Figure 4.3. Dodge small spherical cap of 0.5 degree radius

Table 5.2 Station geoidal undulation

				OSU 86 F/.5	
Station	Gem/10/400	OSU 86 F	NGS (Global)	(L) Local Undulation	L+G
Dodge	-30.41	-30.42	-29.26	-1.52	-30.78
Hardy	-28.15	-28.22	-27.52	-1.13	-28.65
Humbolta	-28.14	-28.17	-28/49	-1.16	-28.65

Table 5.3 Stations orthometric height

Station	NGS (Levelling)	Navcore + OSU 86 F	Navcore + L+G	Geolab Ashtech (G)	Geolab (G+L)
Dodge	307.23	322.66	323.03	307.23(fixed)	308.75
Hardy	357.47	354.08	354.51	356.45	357.58
Humbolta	348.29	346.43	346.91	343.97	345.13

(gravity) maps for each station. See Figure 4.3 for a typical map. The method of spherical cap construction compartmentalizes the residual gravity data points into high and low anomalies and areas of evenly spaced data. The concentrically bound compartments of the spherical cap are also bound by lines of equal azimuth. It is necessary that at least one data point be contained in each compartment and that all compartments fill the circle of the spherical cap.

With the geometry of the spherical cap known, it is entered along with the residual gravity data into the Stokes' formula integration program. This VAX computer program computes the local undulation, or improvement to the global geoidal undulation.

5.0. HEIGHTS ANALYSIS AND CONCLUSIONS

Table 4.1 gives the stations' GPS geodetic heights and single observation standard deviations. Total geoidal undulations result from the algebraic superposition (addition) of global and local geoidal undulation components, transforming these geodetic heights into orthometric heights. These totals for each station are shown in Table 5.2, except for the "OSU86F" column where that model alone provides the total undulation. The "GEM-10/400" column refers to the first method, or Fury model (3) which utilizes the GEM-10 global model improved by a 400 square degree cap. The method of this investigation is abbreviated "OSU86F/.5"; its undulations appear to be greater in magnitude than the others. This may occur because the most gravity data were sensed in this model.

Table 5.1 Station GPS heights

Station	Navcore ellipsoidal height		Geolab Output Using Ashtech observation	
Dodge	292.24	10 m	277.81	(fixed)
Hardy	325.86	14 m	328.93	0.02
Humbolta	318.26	12 m	316.48	0.02

Table 5.4 Orthometric height differences

Line	NGS (Levelling)	Ashtech (G&L)	Distance
Dodge - Hardy	50.24	50.83	129 km
Dodge - Humbolta	41.06	36.38	120 km
Hardy - Humbolta	9.18	12.45	15 km

Table 5.5 Ashtech Loop Misclosure

```
Town    - Hardy     -   43.9575
Hardy   - Humbolta  -  -12.4719
Humbolta - Town     -  -31.6131
                       ─────────
                        -0.1275
```

The total undulations are subtracted from the GPS geodetic heights of each station to obtain orthometric heights (Table 5.3). It appears stand alone (Navcore) determination of the elevation of Dodge has an error of about 15 m. The local undulation improves the elevation determination Hardy and Humbolta by both Ashtech and Navcore by about 0.5 meters. The loop misclosure (See Table 4) using Ashtech indicates that the NGS elevation of Humbolt may be in error by about 3 meters. The orthometric height difference (Table 5.5) indicates a difference of 0.6 m between Levelly and GPS levelling using Global and Local correction over a distance of 120 Km.

It can be concluded that local undulation improves height determination. GPS differential levelling with Local and Global undulation correction can yield an accuracy of 0.6 m per 130 Km or 1 mm per 200 m. The process of determining local undulation by this method is time consuming which could be improved by better programming techniques.

BIBLIOGRAPHY

1. ACSM-NGS. Coordinate Transformation Workshop. Unpublished Xeroxed paper. ACSM, Falls Church, Va., 1987.

2. C.E. Ewing and M. M. Mitchell. Introduction to Geodesy. New York: Elsevier, 1979.

3. R. J. Fury. "Prediction of Deflections of the Vertical by Gravimetric Methods." NOAA Technical Report NOS NGS 28, 1984.

4. R. J. Fury. National Geodetic Survey. U. S. Dept. Commerce, Rockville, Md., Personal communication, 1988.

5. J.D. Gigierano. Geological Survey Bureau, Iowa City, Ia., Personal communication, 1988.

6. W. A. Heiskanan and H. Moritz. Physical Geodesy. Graaz, Austria: Institute of Physical Geodesy, Technical University, 1979.

7. K. Jeyapalan. "Use of Global Positioning System (GPS) for Precise Relative Positioning and Land Surveying." Final Report. College of Engineering, Iowa State University, Ames, Ia., July, 1987.

8. A. H. W. Kearsley. "Tests on the Recovery of Precise Geoid Height Differences from Gravimetry." Journal of Geophysical Research, 93, no. B6 (June 10, 1988): 6559-6570.

9. H Moritz. "Geodetic Reference System 1980." Bulletin Geodesique, 54, no. 3 (1980): 395-405.

10. National Geodetic Survey. Geodetic Glossary. Rockville, Md.: U. S. Government Printing Office, 1986.

11. R. H. Rapp and J. Y. Cruz. "Spherical Harmonic Expansions of the Earth's Gravitational Potential to Degree 360 Using 30' Mean Anomalies." Report no. 376. Dept. of Geodetic Science and Surveying, The Ohio State University, Columbus, Oh., December, 1986.

12. R. H. Rapp. Dept. of Geodetic Science and Surveying, The Ohio State University, Columbus, Oh., Personal communication, 1988.

13. J. Reilly. Surveying with GPS. Unpublished Xeroxed paper. Presented at ASCE/ICEA Surveying Conf., Ames, Ia. ASCE/ICEA, New York, 1988.

14. W. E. Strange, S. F. Vincent, R. H. Berry, and J. G. Marsh. "Detailed Gravimetric Geoid for the United States." In _The Use of Artificial Satellites for Geodesy_. Geophysical monograph 15. Eds. S. W. Henriksen, A. Mancicni, B. H. Chovitz. Washington: American Geophysical Union, 1972. 169-176.

15. W. Torge. _Geodesy_. Berlin: DeGruyter, 1980.

THE RECEIVERSHIP SOLUTION: PROTECTING THE RIGHTS OF
MINORITY STOCKHOLDERS

DONALD J. KAUFMAN, P.L.S.
OLSON-KAUFMAN, INC.
702 5TH AVE. S.
DEVILS LAKE, N.D. 58301

BIOGRAPHICAL SKETCH

The author, Donald J. Kaufman, graduated from Lake Region Junior College, Devils Lake, N.D. and has 28 years of experience in the surveying field, the last 15 of which have been in private practice. He is a charter member of the North Dakota Society of Professional Land Surveyors and is currently president of that organization. A licensed professional land surveyor in four midwestern and western states, Mr. Kaufman is presently vice-president of Olson-Kaufman, Inc. in Devils Lake, N.D.

ABSTRACT

Minority stockholders in a closely-held corporation have long been forced to bend to the will of the majority. Outgunned and outvoted, the minority often have little influence in the day-to-day operations of the firm, yet are held equally liable for many of the corporate actions and debts. Minnesota and North Dakota have developed innovative statutes in recent years to protect the rights of the minority stockholder and to provide remedies in corporate disputes. These statutes consider the fiduciary duty of the majority stockholder to act in a fair and reasonable manner in the operation of a closely-held corporation. This paper examines the options available to oppressed minority shareholders.

INTRODUCTION

The start-up of a small company can often be likened to a marriage. The principals are usually friends entering optimistically into the venture without giving much thought to corporate charters and by-laws detailing respective rights in situations such as dissension and dispute. The partners feel they will always have an amiable relationship, and there is no need for heavily lawyered contracts.
Unfortunately, like many marriages, the partners forget the reality that life can often be cruel and unforgiving; that human nature cannot be accounted for, and that eventually the honeymoon will be over and the courts may well have to decide their fate.

Corporate Structure

The small engineering-surveying firm typically consists of two principals and less than twenty employees. The principals are usually the founders and corporate officers holding all the issued stock except for a few shares exchanged for organization expenses incurred by lawyers or accountants.

After the initial organization, the principals usually tend to go on with their respective careers, lapsing into the traditional roles that have structured these firms for decades. The engineer designs, markets and manages the office. The surveyor/technician coordinates the survey crews (perhaps runs a crew himself), helps design and manages projects or outside personnel. In the ideal situation, this combination should work efficiently and does in many cases. However, having such a defined split in the management roles with no checks or balances leaves the door open to abuses in the financial management by the majority stockholder, who is also typically the engineer and CEO

Although expected to do well in this capacity, engineers as a whole are notoriously bad managers. Their education and training does not prepare them to administer the business of the corporation, and most tend to do so by the "seat of their pants". Hence the 'boom and bust' cycles common to engineering firms that result from the erratic nature of the work, as well as the unsophisticated management and marketing skills of the CEO.

The problems are often compounded because neither principal is well-trained in management skills, yet the corporation does not have sufficient resources to hire an outside manager. An imbalance is created by placing the direction of the corporation in the hands of one person, thereby isolating the minority from many important financial decisions.

A CORPORATION IN TROUBLE

Place yourself in the position of a minority shareholder in a small, closely-held corporation. The remaining stock is in the hands of one majority owner. There are no other shareholders. The office staff consists of one secretary/office manager who does billing and keeps the books. An independent accountant does the taxes and annual financial statements. Your share of responsibilities keeps you in the field most of the time so the daily financial decisions are handled by the majority shareholder. You receive a monthly financial statement, which you probably can't understand very well. But the bills are getting paid, your salary is paid on time and the jobs keep rolling in regularly and major financial decisions are shared.

Eventually, however, human nature may demonstrate its darker side, as the embezzler mentality surfaces with the partner skimming a few dollars extra, intending, of course, to pay it back someday. Don't get me wrong, this is not your typical loan to an officer, which is a normal corporate practice. This is usually cash needed by the CEO to pay for something he may be embarrassed to disclose, or maybe he has acquired a girlfriend somewhere, or his wife wants a new car, or he wants a new boat, or he's a little short for his children's college tuition. The list goes on and on. But inevitably, once it starts, unauthorized 'perks' tend to snowball rapidly.

Not to worry, right?

You have your checks and balances in place and you're getting a financial statement every month and it makes you feel secure. You also have your office manager to alert you to the fact that maybe something is wrong with the financial management or eventually, the accountant will call you all together at year's end and say "we have a problem here". Wrong! In our scenario, the office manager is not going to react until the problem is too far out of control or has a "don't get in the middle" attitude from the beginning. Perhaps the accountant is only getting the information that the CEO wants him to get, or maybe they're buddies and he's helping to cover the trail. The 'old boy' attitude has typically been to look at some misappropriation of corporate funds as just another perk that the majority shareholder was entitled to, and not a serious situation requiring confrontation. However, you will note that the word misappropriate has the same meaning as defalcation, stealing or embezzling.

Someone once said that a democracy can last only as long as it takes for the general public to realize that they can vote themselves largess out of the public treasury. The same concept applies directly to the business world. We have already mentioned that engineering has traditionally been considered a 'boom or bust' industry, usually because of the CEO's inept management style. So let's assume that the majority shareholder skims funds during the peak years, can't replace them during the inevitable lulls and eventually forces the corporation to use its line of credit. This means putting your name on the dotted line if you want to continue receiving that regular paycheck. The bank doesn't care how many shares of stock you own, when you personally guarantee a loan, you are as fully responsible to pay it back as the CEO. Take a worse case scenario, and say that the corporation is in such disperate straits that the payroll taxes are not being deposited regularly. You, as a corporate officer, are 100% personally liable to the IRS not only for the payment of these taxes, but for all the interest and penalties that accrue. If your CEO decides to take a vacation when the IRS comes to call, they could care less as to how many shares you own or what you know about financial statements...they will hold you responsible. The most exaggerated business myth that exists today is that of corporate immunity. The corporate veil, as it is commonly described, has been pierced by the courts so often over the years so as to make it virtually nonexistent. So don't ever let yourself forget that you could be held responsible for the actions of the majority stockholder whether you are aware of them or not.

This is a pretty bleak situation that, unfortunately, happens all too often. As the minority shareholder its up to you to protect yourself. The majority is not going to voluntarily give up any control, so you're going to have to stick to your guns until you get a workable management plan you both can live with. This is best done before you go into business or can be implemented along the way if things haven't completely unraveled as yet. However, if you can't get an effective plan in place, it may be better to cut your losses and walk away from a potentially bad situation if possible. An exception exists where you can't escape some liability simply by resigning. So if you are unlucky enough to be saddled with an unscrupulous or incompetent partner, you are eventually going to have to face the grim reality that the firm may well be a sinking ship and you are going to go down with it. Hopefully, this revelation will come before the situation forces you into personal bankruptcy.

Minority Rules

Fortunately, the courts are addressing the rights of the minority stockholder in small closely-held corporations. States such as North Dakota, Minnesota, California, among others, have statutes granting ill-treated minority shareholders sufficient powers to dissolve a corporation. Although these statutes are not so well defined everywhere, all states have laws that address the irresponsible behavior of majority shareholders and other remedies in the case of oppression or mistreatment. The principle behind all these statutes is that the majority shareholder has a fiduciary duty to treat the minority shareholder equitably.

The corporate majority was at one time generally considered to be a sacred entity. However the courts are now ruling more frequently that every shareholder has a right to be treated fairly and if not, they will not infrequently side with the minority shareholder.

If you as the minority shareholder suspect an abuse of the corporation's financial resources is occurring at the hands of the majority shareholder, your immediate concern should be to get as much documentation of the suspected transactions as possible. As a shareholder you have access to all company records. Hire a private accountant if necessary to go through the books for you, if you don't have the expertise to do so yourself. Take this information to an attorney who specializes in corporate law and he will advise you as to whether or not you have a case. Usually the first consultation is free in situations such as this.

Assuming you have a case and decide to proceed, the attorney will present the evidence to the District Court in the form of a petition. The petition will request that the court appoint a receiver or trustee to oversee and protect the best interests of the creditors and all the shareholders in the corporation. The Judge will issue an order to show cause, setting a date (usually within 2 weeks) so that the defendant may be heard as to why a receiver should not be appointed. Normally the majority won't fight this because, if you have done your homework, the Judge will rule in your favor on the basis of the evidence and will appoint a receiver of his choice.

Eventually the parties or their attorneys will come to an agreement as to who should act as receiver. This person should be a disinterested third party, usually a CPA or someone knowledgable in financial matters, such as a banker. This person must be bonded and usually appointed for a 6 month period. The fees charged by the receiver as well as those of the attorneys should be paid for out of corporate funds since this action is considered to be in the best interest of the corporation.

The receiver should immediately take control of the corporate assets, and no checks written or contracts entered into without his/her approval. Some receivers actually write all of the checks and sign all of the contracts during their appointment and it's probably a good idea for them to do so.

The appointment should effectively stop the abuses from continuing and if the corporation is salvagable, it will be back on its feet in a relatively short period of time. The receiver has all of the power of the court to negotiate with creditors on behalf of the corporation and is usually successful in extending terms on outstanding debts.

During this six month period, the receiver will determine the viability of the corporation and make recommendations to the court. If the shareholders cannot reach an agreement during this period, or if the concerns of major creditors haven't been resolved, the Judge will eventually have to make a ruling.

Depending on the attitudes of the shareholders and the types of abuses which have occurred, the Court has the option of utilizing one of the following remedies: 1. The court can require that the receiver continue for another six month period or appoint a new receiver if there is some hope that the differences can be resolved during that time. 2. The court itself can retain Jurisdiction in lieu of a receiver. 3. Cancelling the issued stock and ordering a 50/50 balance of ownership. 4. Permitting the minority to purchase an increased interest in the firm. 5. Award damages to the minority for injuries suffered because of oppressive conduct. 6. Declare a dividend to the injured stockholder. 7. Involuntary dissolution.

If the corporation is in a precarious financial position, the Judge will probably cause it to be dissolved as per the statutes of the particular state. The creditors and injured shareholder would then be compensated with the remaining assets in the corporation.

On the other hand, if the corporation is still a viable entity, the Judge may appoint a panel to appraise the firm and force one party to buy the other out. This does not necessarily end the business but simply ends the relationship between a particular group of shareholders. At this point, the corporation may also be put up for bids and any party, including the minority can bid on it.

The message here is that the corporation and the majority shareholder are no longer inviolate parties in the eyes of the court.

Management Planning

Preventative measures are always less painful than the cures listed above, so the corporation should be structured from the outset around solid management skills. Nobody can "shoot from the hip" very long without getting into serious trouble eventually. You probably educated yourself as a surveyor, now is the time to get a financial education. There are many good books geared for the layman on financial management available. Start reading and continue reading for the rest of your life. Initially you may find this pursuit dull and uninteresting, but it's a lot less taxing than sitting across the courtroom from your partner and his attorney. Nothing, of course, can guarantee safety from a courtroom battle with your partner, but an education will give you an edge by helping you spot trouble before it has gotten completely out of hand.

Learn to read a financial statement and see to it that all financial decisions, no matter how trvial, are brought to the attention of all the shareholders. Although inconvenient, it may be a good idea to have two signatures on all company checks. At least make sure that all corporate expenditures are reviewed by you and explained to you. Another excellent idea often overlooked, is to have the company bank statement reconciled monthly by a third party (often banks will do this as a service with a modest charge).

The accountant and office manager should be informed in no uncertain terms that you expect to be notified if anything, no matter how trivial, seems amiss in the corporate finances. Don't expect any real help from the office manager because they are just another employee and usually don't care to get caught in the middle of any management disputes. The accountant, on the other hand, has an ethical duty to inform all the shareholders if anything seems amiss. Also insist the corporation use the accrual rather than cash basis accounting for its books. This system helps match expenses with revenues on ongoing projects, and makes it easier to spot mismanagement early in many cases.

Unfortunately, there is no foolproof system to prevent misappropriation of corporate funds and all of the checks and balances you can device won't prevent an unscrupulous partner from trying to get a bit ahead at your expense. At least the courts are beginning to recognize the scope of the problem in small closely-held firms and offer some types of relief for the minority shareholder.

CONCLUSION

These are perilous days for the professional surveyor,

particularly as a shareholder in an engineering company. We are faced with unending liabilities with every survey or project undertaken, and we shouldn't be forced to shoulder liabilities that are not of our own creation.

Long considered a second class professional, the surveyor must again be allowed to take control of his destiny as a viable part of the corporate management team, no matter what percent interest he holds in the company.

The courts should be the last resort for any corporate disputes, because it will likely sever or forever taint the relationship of the shareholders. However, if the cavalier management style of the majority shareholder has placed you in a position of liability beyond your control, the only answer may be a court-appointed receiver.

REFERENCES

Hood, Kurtz and Shores. Closely Held Corporations in Business and Estate Planning. Vol. I, 1982, Little & Brown.

O'Neal, F. Oppression of Minority Shareholders. 1975, Callaghan & Co.

O'Neal, F. and Derwin, J. Expulsion or oppression of Business Associates. 1971, Duke University Press.

AUTOMATED ENHANCEMENT OF SHAPES OF REGISTRATION PARCELS

A. Krupnik and B. Shmutter

Department of Geodesy
Technion - Israel Institute of Technology
Technion City, Haifa 32000 ISRAEL

ABSTRACT

Procedures for automated retrieval of the layout of parcels from data obtained by scanning and vectorization of a cadastral map have been described by the authors in previous publications. The result of that process, nodes and arcs, although representing the boundaries of the parcels correctly, are still incomplete in two respects: *(i)* Boundary lines are represented by excessive numbers of nodes, which is not the case in the original map, and *(ii)* Shapes of lines are not always preserved. To complete the above process, further editing routines have been developed which correct these deficiencies: Identification of strings of points constituting a boundary line shared by a number of neighboring parcels; Fitting straight lines to strings of points when it is found that they should represent such lines; Determining for each node a single pair of coordinates, in case it appears more than once in the data; Rearranging the layout of the parcels according to the final set of nodes. As a result of that process superfluous data are eliminated, each parcel is represented by genuine nodes only - its corners, and the graphical quality of the output is enhanced considerably. An exposition of the procedures and corresponding examples are given in the paper.

INTRODUCTION

Data acquired by digitizing cadastral maps usually consist of nodes and arcs. Because of inaccuracies inherent in the digitizing process, the layout of the parcels as described by the digitized data becomes slightly distorted with respect to its original. Thus editing operations are needed to compensate for such distortions. Procedures for adjusting the cadastral data captured by manual digitization are discussed in Shmutter & Doytsher (1990). Additional problems arise when the cadastral data are acquired by scanning the map with subsequent vectorization and automated extraction of boundaries, as described in Krupnik & Shmutter (1989). The data yielded by that process are again nodes and arcs, and at the top of it lists of contiguous arcs forming the boundaries of the parcels. Unlike the data obtained by manual digitization, where each side of a parcel is defined by two nodes (in case it is a straight line) and the entire parcel by its corners only, the automated process provides parcels described by excessive amounts of data. For example, a long string of points which should represent a straight line, and as such be determined by its two extremities only, is actually represented by a sequence of short segments (fig. 1a). Besides, intersections of lines are distorted by the raster to vector conversion (fig. 1b and 1c). Another example of the data redundancy is depicted in fig. 2. It follows that the process has to be complemented by additional editing routines to enhance the shapes of the boundary lines on the one hand and to reduce the superfluous data on the other.

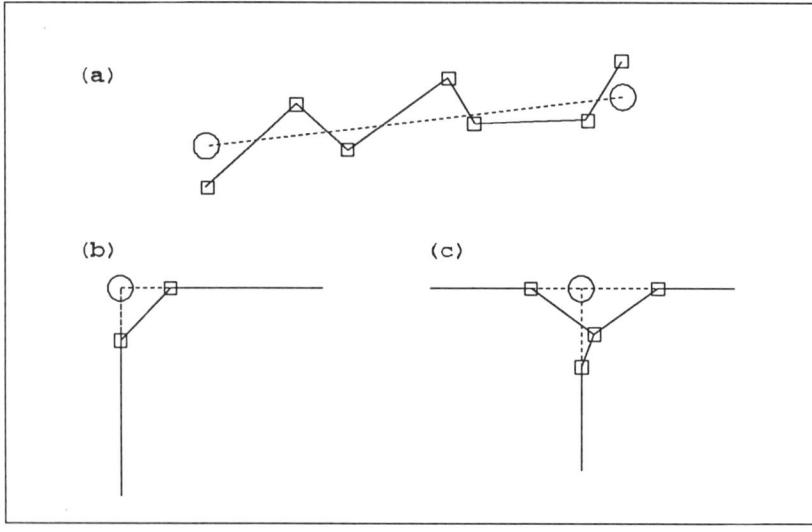

Figure 1: Deficiencies inherent in the data
(Exaggerated for graphic display reasons)

TERMINOLOGY

The automated process described in Krupnik & Shmutter (1989) uses a data structure in the form of a graph. In this connection it is necessary to explain a few terms which are used in the following:

A node: an extremity of a line segment.
n-d node: a node common to 3 or more segments.
"circle" node: a node generated at the center of a circle which represents a survey mark on the map.
Internal nodes (or 2-d nodes): common to two nodes only.
Neighbors: nodes i and j are neighbors if there exists a line segment between them.

EXPOSITION OF THE EDITING ROUTINES

<u>Identifying strings of segments which should constitute a straight line</u>

A sequence of short segments is said to represent a straight line when it meets the following conditions:

1. All its nodes (except the first and the last) are internal and none is a circle node.
2. The offset of each node from the straight line connecting the two extreme nodes is less than a predefined value.

The identification starts with a resampling of the data aimed at establishing sequences of segments, the end points of which are either n-d nodes or circle nodes (first condition). Each of such sequences when found is examined whether it satisfies the second condition. If that is not the case, it is subdivided into sub-sequences to meet the above condition.

<u>Concatenating segments to generate frontages of parcels</u>

A sequence of short segments constitutes a frontage of a single parcel when it starts and terminates at n-d nodes, or 2-d nodes at which the incident segments form an angle deviating considerably from two right angles.

On many occasions sequences of segments represent a long frontage common to a number of neighboring parcels. Identifying such a line is effected by the following procedure:

1. Starting with a sequence of segments (identified in the previous stage as a straight line) chosen at random the procedure tries to locate another sequence which shares one of its end nodes with the initial sequence.
2. The offset of the common node from the line connecting the extreme nodes of the sequences located before, and the current sequence, is computed. If it does not exceed a given magnitude it is concluded that the sequences so found are pertaining to one straight line.
3. The following sequences are located in a like manner. The end node of the last sequence is declared as the end node of the line being assembled.
4. The first sequence chosen and the strings of sequences found on its both ends establish a straight frontage.

That procedure is repeated as many times as there are frontages to be concatenated.

It should be mentioned that in some instances a single segment is by itself a frontage of a parcel.

Fitting straight lines to strings of nodes

A least squares algorithm is used to fit a straight line to each string of nodes. These differ from one another with regard to the reliability of their locations. Therefore various weights are introduced in the adjustment process. The highest weight is assigned to a circle node, since its location was found to be most reliable. The lowest is assigned to an n-d node, because the intersection of lines producing such a node is distorted to some extent by the raster to vector conversion. An intermediate weight is assigned to the other nodes.

The adjustment provides corrected coordinates of nodes satisfying an equation of a straight line. Most of these nodes are irrelevant for further processing and may be deleted. Only those which are corners of parcels (referred to in the following as major nodes, or majors) are retained. The majors are identified as n-d nodes positioned on the adjusted line, as well as its two end nodes. The adjusted coordinates of the majors and their sequential numbers are stored. The latter are required for relating major nodes common to two or more lines.

Determining coordinates of major nodes

In the initial data each node was determined by a single pair of coordinates, regardless of the number of segments it was related to. That situation is not necessarily the case now, since the fitting of straight lines, being carried out for each line separately, may assign to one and the same major node different coordinates depending on the adjusted lines on which it is positioned.

In order to properly arrange the layout of the parcels it is mandatory to define unique coordinates for each major. The majors fall under two classes: *(i)* major nodes related to three straight lines at least, *(ii)* majors related to two lines only. The position of a node from the first class results from an intersection of the lines it pertains to, performed by a least squares adjustment. This alters the positions of the lines taking part in it, which in turn has an effect on the locations of other major nodes lying on those lines. Consequently, it becomes necessary to reiterate the adjustments of the majors until they assume their final locations. Final positions of majors of the second class are established by intersecting the lines they pertain to. Since no adjustment of coordinates is required here, the pairs of lines participating in the intersections maintain their locations, thus this step does not necessitate any iterations.

The outcome of the above procedures is illustrated by fig. 3. Comparing figures 3 and 2 shows clearly that the parcels become defined by genuine corners only, and the superfluous data are being deleted.

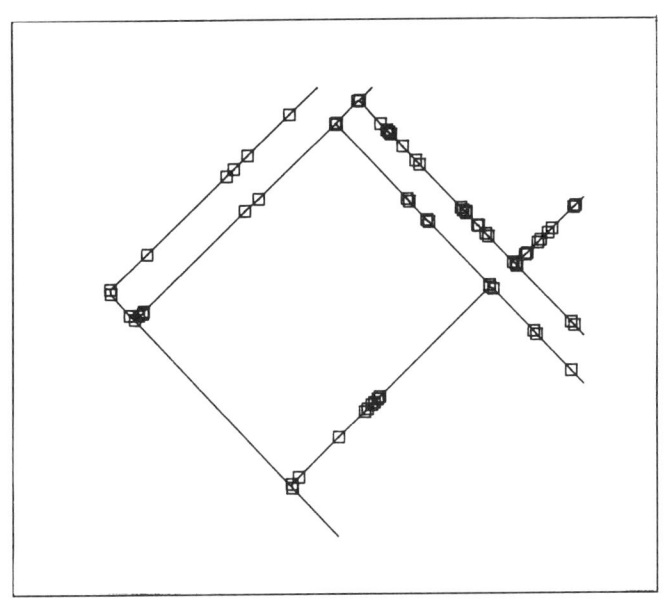

Figure 2: Data redundancy

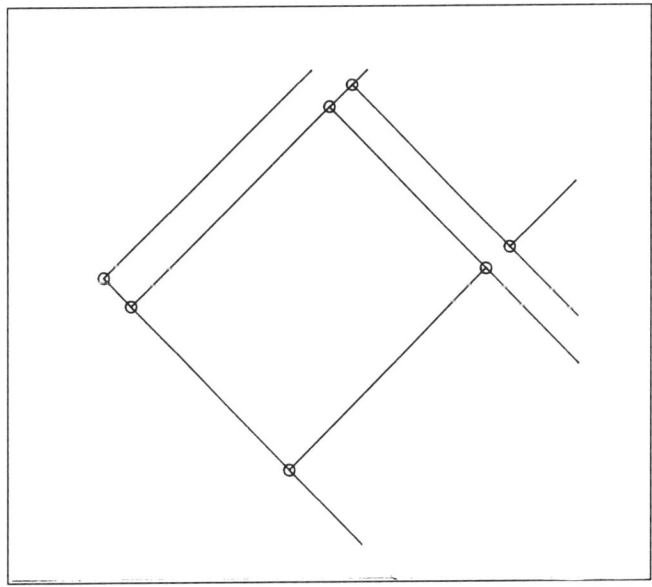

Figure 3: Outcome of the process

Rearranging the layout of the parcels

The former steps yield data differing from the initial data in that they consist of major nodes only. Owing to the fact that many of the initial (redundant) nodes have been deleted it is now necessary to rearrange the data in order to define the layout of the parcels anew, which requires to assign to each major node new neighbors.

Having arranged proper lists of neighbors, the parcels are now assembled according to the algorithm described in Shmutter & Doytsher (1990).

SUMMARY

Data derived from a cadastral map by scanning, vectorization and automated editing are further enhanced to produce a layout of parcels, which preserves the straightness of lines and its congruity with the original map.

The enhanced layout of the parcels is suitable for being incorporated into a GIS as a component of a cadastral layer. As the parcels are described by strings of genuine corners only, they can be conveniently utilized for various applications.

It should be noted, that the process is fully automated and does not need any manual assistance.

REFERENCES

Shmutter, B. & Doytsher, Y. (1990): "Assembling Closed Polygons". Survey Review, Vol.30:235, pp.109-220.

Krupnik, A. & Shmutter, B. (1989): "Automatic Recognition of Property Lots from Old Cadastral Maps". Presented at the 14th ICA Conference, Budapest.

SYSTEMATIC EFFECTS IN SINGLE-FREQUENCY GPS IN THE KINEMATIC MODE

Alfred Leick
Quanjiang Liu
Earl Burkholder

University of Maine
Orono, ME 04469

ABSTRACT

Single-frequency GPS relative positioning is effected by residual components of the ionosphere, the troposphere, multipath, and geometry of satellite constellation. There are other error sources as well. In static application the impact of these effects is hopefully minimized by observing over a long period of time. In kinematic applications, particularly continuous kinematic GPS, the positions of the roving antenna are determined each observation epoch. Several tests cases are studied to determine and quantify the impact of systematic errors on the position of the roving antenna. Baselines of various lengths are analyzed. This includes cases where the "roving antenna" actually does not move (static) and cases where the roving antenna is mounted on a platform to monitor 9 feet of tides along the coast of Maine.

THE PUBLIC DOMAIN LAND TENURE SYSTEM IN THE UNITED STATES

MARLIN LIVERMORE
UNITED STATES BUREAU OF LAND MANAGEMENT
2850 YOUNGFIELD STREET
LAKEWOOD, COLORADO 80225

BIOGRAPHICAL SKETCH

Marlin (Lin) Livermore is a cadastral surveyor employed by the U. S. Bureau of Land Management, Colorado State Office. During his 30 year career with the Bureau his assignments have included serving as a Chief of Parties and as a supervisor/manager in various cadastral offices. As a cadastral surveyor Lin has executed surveys and resurveys of the Public Land Survey System in 13 of the western and mid-western states. Lin has been licensed to practice land surveying in New Mexico since 1965.

ABSTRACT

A dual land tenure system exists in the thirty public land states consisting of a federal and the state and local government systems. Surveyors in these states are susceptible to performing surveys under the rules of each system and therefore must know a dual set of survey rules. The Manual of Surveying Instructions, 1973, (Manual) define the rules of survey for the federal system. Proper application of the Manual rules requires knowing the definition of "manual terms" and how to distinguish between and interpret the various mandated, general and specific rules found in the Manual.

INTRODUCTION

A two tier land tenure system exists over approximately eighty percent of the United States. The federal tier consists of its official public records and surveys which includes the establishment of the Public Land Survey System. The second tier consists of state and local land records and surveys executed by licensed professional surveyors.

In the thirty public land states all title to the lands originated from the federal government. The legal descriptions and positions of the lands are defined by the Public Land Survey System when the federal government transfers title. Once the lands have left federal ownership legal jurisdiction over them shifts from the federal government to the state and local governments.

The United States' Bureau of Land Management maintains the federal land tenure records and has sole congressional authority to execute official federal surveys and make rules and regulations on how federal boundaries are to be surveyed.

SURVEYORS IN THE PUBLIC LAND STATES PERFORM SURVEYS UNDER BOTH SYSTEMS

SURVEYORS WORKING IN THE PUBLIC LAND STATES MUST KNOW THE RULES OF SURVEY FOR BOTH SYSTEMS.

It is essential that all surveyors in the public land states understand the existence of the dual land tenure system and the basic authorities and laws under which each tier operates. The two tiers have different rules on who has authority to execute surveys and on the methods and procedures used in their execution.

The location of federal boundaries fall under federal jurisdiction and proper survey procedures for the location of these boundaries are defined by federal law. Legal disputes affecting federal ownership of property are generally decided in federal courts.

In contrast the survey procedures for the location of boundaries that fall under state jurisdictions are defined by state laws and legal disputes on the ownership of private property are generally decided in the state court systems.

State licensed surveyors are required to apply the rules applicable to the federal tier when:

    1. A boundary survey affects federal ownership.

    2. State authority has adopted federal surveying rules.

The extent federal rules has been adopted by states varies from state to state, however almost all, if not all, public land states have adopted the federal rules to some extent through case or statutory law. For instance, New Mexico state law provides in its Engineering and Surveying Practice Act that:

    Section 29. RESTORATION OR REESTABLISHMENT OF MONUMENTS--STANDARDS AND PROCEDURES.--

    a. When any registered surveyor in the course of surveying is required to use any lost or obliterated survey monument that has originally been established by a federal agency, the registered surveyor shall restore or reestablish such survey monument in compliance with the standards and procedures set forth in the "manual of surveying instructions for the survey of public lands of the United States."

The federal government adopts state survey rules in certain situations. The official federal surveys of lands reacquired by the federal government are based on state survey rules. The federal government recognizes and applies state rules on certain riparian survey issues.

THIS PRESENTATION IS ABOUT FEDERAL SURVEY RULES.

This presentation is based on the federal rules of survey. The thesis and opinions introduced in the presentation are not always applicable to state rules of surveying.

THE SOURCE OF FEDERAL SURVEY AUTHORITY AND RULES

THE UNITED STATES CONSTITUTION.

The source of the federal survey authority is the United States Constitution. The constitution created spheres of powers for the federal government reserving all other powers to the state governments and the people. One sphere of federal power created by the Constitution was:

> "The Congress shall have power to dispose of and make all needful rules and regulations respecting the territory or other property belonging to the United States." U.S. Const. art. IV, sec. 3. Based on this power the United States Supreme Court has held: "(T)he courts will not pass on how congress' trust over public land shall be administered." <u>Alabama v. Texas</u>, 347 U.S. 272 (1952). (Choper.)

"This Constitution, and the Laws of the United States which shall be made in Pursuance thereof...shall be the supreme Law of the Land; and the Judges in every State shall be bound thereby, any Thing in the Constitution or Laws of any State to the Contrary notwithstanding." Id., art. VI. (Note: Capitalization as it was originally written.)

The authority of Congress over the public lands is plenary. Not only does congress have the power to make rules concerning the survey of federal lands, no state laws, rules or regulations can usurp this authority. Surveys of federal boundaries must be executed under federal rules.

THE SOURCE OF THE BUREAU OF LAND MANAGEMENT'S SOLE OFFICIAL FEDERAL SURVEY AUTHORITY.

Congress has delegated the authority to execute official surveys of the federal lands to the Secretary of Interior, who has designated the Bureau of Land Management to perform the duties:

"The Secretary of the Interior or such officer as he may designate shall perform all executive duties appertaining to the surveying and sale of the public lands of the United States..." 43 U.S.C. 2. The Bureau of Land Management has sole official survey authority on Indian Lands under 25 U.S.C. 176. These authorities have been interpreted to include the authority to make rules and regulations concerning the survey of all Indian and federal lands. These rules and regulations are outlined in the Manual.

THE MANUAL OF SURVEYING INSTRUCTIONS DESCRIBES HOW FEDERAL LANDS SHOULD BE SURVEYED.

"The Manual of Surveying Instructions describes how cadastral surveys of the public lands are made in conformance to statutory law and its judicial interpretation." Manual, sec. 1-1.

The various Manuals of Surveying Instructions, the current being, the Manual of Surveying Instructions, 1973, plays a primary role in the surveys of the public land states. This "technical bulletin" is recognized by the federal and state courts as providing the proper rules for the survey of the federal lands and in many cases private lands.

APPLICATION OF THE MANUAL TO FEDERAL SURVEYS

The improper application of survey rules is a frequent source of litigation in the surveying profession. This is true whether state or federal rules are applicable. These types of survey errors are created when surveyors do not know the existence of rules and therefore do not realize they need to search for them.

Surveyors are susceptible to misinterpretation of the Manual if they lack a general knowledge of its contents or the definition of terms or apply the rules out of context, i.e., applying rules relevant to original survey to resurveys or vice-a-versa.

PROPER SURVEY PROCEDURES REQUIRE DEFINING THE FACTUAL SITUATION AND APPLYING THE RELEVANT RULES.

The proper application of survey methods and procedures for any given situations, whether applying Manual rules or rules established by state authority requires, determining the factual situation and based on the factual situation: (1) Identifying all relevant rules. (2) Complying with all mandated rules. (3) Applying the relevant general and/or specific rules.

MANUAL TERMS ARE MEANT TO MEAN WHAT THEY ARE CHOSE TO MEAN, NEITHER MORE OR LESS.

Interpretation of the Manual requires knowing the definition of "manual terms".

"When I use a word", he said scornfully, "It means just what I choose it to mean, neither more or less." Stated by Humpty Dumpty in, Through the Looking Glass. (Carroll).

A surveyor can also be trea ed scornfully if he misinterprets "manual terms" by choosing them to mean more or less than they actually mean.

Example of a definition of a "Manual Term".

Reestablishment of a lost closing corner requires proportioning between the "nearest <u>regular corners</u> to the right and left." Manual, Section 5-41. Emphasis added.

Section 5-32 defines regular corners relative to a standard parallel as, "standard township, section, quarter section, and sixteenth-section corners, and meander corners and also <u>closing corners</u> which were originally established by measurement along the standard line as points from which to start a survey." Emphasis added.

Adding or subtracting from the definition of when a closing corner is or is not a regular corner will result in the erroneous positions of reestablished corners.

<u>MANDATED, GENERAL AND SPECIFIC RULES.</u>

At first glance the Manual appears to contradict itself. The seemingly contradictions disappear when we understand the definition of terms and recognize that the rules fall into three separate categories, which I have designated as mandated, general and specific rules:

MANDATED RULES. Mandated rules applies in all relevant survey situations. An example of a mandated rule is the rule, established by statute that bona fide rights must be protected in the execution of all resurveys. 43 U.S.C. 772.

This rule is relevant to all resurveys and therefore it is mandatory that it be applied on all resurveys. The proper application of any general or specific rule during a resurvey will fall under the umbrella of this rule.

GENERAL RULES. General rules are not narrowly particularized. When relevant, these rules always apply if they are not superceded by particularized criteria defining where a specific rule applies in lieu of a general rule.

The basic proportioning procedures for the reestablishment of lost corners are examples of general rules.

SPECIFIC RULES. Specific rules define procedures to apply when and only when explicit factual conditions are encountered in a survey or resurvey. Once the criteria is met the specific rule is applied in lieu of the general rule.

An example of an specific rule is the acceptance of good faith location in lieu of proportionate measurements. The existing factual conditions applicable to acceptance of good faith locations are stated in sections 6-16, 6-17 and 6-18 of the Manual.

It is incumbent on the surveyor to establish that existing conditions meet the required explicit criteria to justify the use of a specific rule.

APPLICATIONS OF MANDATED, GENERAL AND SPECIFIC RULES.

General rule.

Closing corners are generally not used as control to determine the alignment or to reestablish the position of lost corners on the line they close on. ID. 5-41.

Reason for the rule: An original line is considered fixed in position. A closing corner established subsequent to when a line was originally established cannot move the position of that line.

In addition, closing corners are not generally used to proportion lost corners on the line that they close upon because the original one-way ties made from closing corners to the nearest adjacent regular corners have often been found to be inaccurate in both alignments and measurements.

Specific Rule

"Where there has been extensive loss of corners, particularly the senior corners, the existent junior corners may also constitute the best available evidence of the line itself. In such a case they will exercise control for both measurement and alinement." ID. Section 5-35. Emphasis added. The proper interpretation of this rule requires knowing when a closing corner is defined as a junior corner.

Reason for the specific rule: The junior corners are considered the best available evidence of the position of the original line when no evidence of the senior corners or survey remains.

Note how the basic relationships between general and specific rule applies in the above example. The general rules apply unless the criteria for the specific rule is met. Careful, there are other specific rules that provide an exception to the general rule used in this example. For instance, the definition of a regular corner creates a specific rule for the use of closing corners as maximum control for proportioning.

The next example further demonstrates the relationships between mandated, general and specific rules.

PROPORTIONING.

When to proportion and when not to proportion to reestablish the position of lost corners is a common problem we encounter in our surveys. If we are surveying federal boundaries criteria have been established to assist in making these decisions. If we are surveying under state authority the federal rules may or may not apply, dependent on the various states case and statutory laws.

What rules apply when your procedures or measurements disagree with the position of another surveyors reestablishment of a lost corner?

Mandated Rule: All resurveys shall protect bona fide rights of land owner.

Reason: Federal law. 43 U.S.C. 772. Protection of property rights protected under the Fifth Amendment of the Constitution.

General Rule: After all original and collateral evidence has been developed proportionate measurement will always be employed to relocate a lost corner unless outweighed by conclusive evidence of the original survey. ID. 5-21

Reason: Proportionate measurement harmonizes surveying practice with legal and equitable considerations. ID. 5-21.

Through equitably dividing acreage the courts have accepted proportionate methods for the restoration of lost corners as protecting bona fide rights.

Specific Rule: "When a local reestablishment of a lost corner has been made by proper methods without gross error and has been officially recorded, it will ordinarily be acceptable. Monuments of unknown origin must be judged on their own merits, but they should never be rejected out of hand without careful study." ID. 6-28.

Criteria for the specific rule: (1) The position of the monument must be determined by proper procedures. and (2) Without gross error. (3) The monument has to be recorded. or (4) If not recorded I would interpret the criteria for acceptance is met if the position of the monument meets the criteria of proper procedures without gross error and was established at a time when state law did not require recordation.

Reason: In my opinion the reason for this specific rule is that the stability of property lines is considered a property right and when the loss of land is "de minimis", (the discrepancy is kept to a minimum by the "gross error" limitation), it is more equitable to protect the stability of property lines than to protect minor adjustments in position or acreage.

If the factual criteria for this specific rule is met it is incumbent on the second surveyor to accept the reestablished position of the first surveyor, if not, the monument is unacceptable under this specific rule.

The position may be acceptable under other specific rules, applicable to the protection of bona fide rights, such as, "good faith location" or specific rules applicable to tracting lands if an original survey has been canceled.

These few examples are a small sample of "terms", and mandated, general and specific rules that exist in the Manual.

CLOSING

During this presentation not once was GPS mentioned, not once was computers mentioned and not once was coordinates mentioned. These TOOLS of the profession are important. I am enthralled with new technology as well as the next surveyor and welcome its contribution to the profession. However, before the GPS or computers are turned on, before coordinates are developed--a Surveyor, qualified in interpreting boundary evidence, competent in applying the surveying rules of both the state and federal tiers of the land tenure system must determine the correct positions of the corners through proper application of the rules of surveying.

BIBLIOGRAPHY

Lewis Carroll, Alice in Wonderland and Through the Looking Glass, Grosset and Dunlop, 1948.

Jesse H. Choper, Constitutional Law, Gilbert Law Summaries, 23rd ed., Harcourt Brace Jovanovich Legal and Professional Publications Inc., Chicago, Il.

Manual of Surveying Instructions, 1973, U.S. Department of Interior, Bureau of Land Management, Technical Bulletin 6.

New Mexico Engineering and Practice Act. NMSA (1978).

Title 43, United States Code.

Title 26, United States Code.

United States Constitution

TOTAL QUALITY DEVELOPMENT IN A LAND SURVEY FIRM

Jim Loy A.R.I.C.S., F.Inst.C.E.S.
Managing Director
Loy Surveys Ltd, Glasgow, U.K.

Loy's quest for BS5750 began with the commissioning of a market research study three years ago.

While the findings were positive the research indicated that some improvements in quality control procedures could be made. At this stage a Quality Assurance form was introduced outlining the technical requirements and client information for each new project.

The initial report by the consultant showed many elements of sound practice within the management system of Loy Surveys. Confidence was placed in the ability of the surveyors to self manage their own projects, including dealing direct with clients and meeting their exact requirements, using the QA form.

The existing informal quality systems required to be documented to the level of current compliance to ensure that the good practices were maintained in an orderly manner. The existing controls were then reviewed against BS5750 and extended as necessary to ensure that all processes operated to the quality principles of BS5750.

TOTAL QUALITY DEVELOPMENT
IN A LAND SURVEY FIRM

Jim Loy A.R.I.C.S., F.Inst.C.E.S.
Managing Director
Loy Surveys Ltd, Glasgow, U.K.

Summary

In this paper the author outlines why Loy Surveys embarked on the road to quality and how they went on to become the first Land Survey firm in the UK to achieve registration to ISO 9000/BS 5750 Quality Management Systems.

The key to change

Change is vital to allow an organisation to grow and prosper. Without change, the organisation withers and dies.

Change is not accomplished easily. You must want change enough, to work hard enough to bring change about. It takes time. It can hurt. Change however is optional - SURVIVAL is not compulsory !

Why Quality

Quality-awareness is on the rise. The interest in quality is growing, all over the world. Customers and users are becoming more and more demanding. They are no longer willing to accept inferior quality. People are insisting that the public sector too improve the quality of its services.

When will private industry and the public sector understand it's time to listen to public demands for quality ?

Loy Surveys Experience

The company was founded in 1980 and has achieved sustained growth through repeat business. Success, we believe is based on giving a better service, not on cost-cutting. Our people have a sense of pride in company achievement this creates a feeling of well being and encourages the development of a positive environment, team spirit and a higher level of personal quality. By the mid 1980's the company's growth had increased to the level where it was evident that the Owner/Manager would not be able to deal personally with all the clients on the technical requirements of each project.

Rather than introduce a management structure at that stage, confidence was placed in the ability of the Surveyors to manage their own projects, including dealing direct with the client. At that time a QA control form was introduced for each project to help document the requirements of each job.
As a result of the decision to increase the responsibility of our Surveyors they gained a great deal more job satisfaction and enjoyed the compliments they received direct from the clients they dealt with.

Why the need to change

During late 1987 we commissioned a market research report to determine how well we were doing and how we could further improve our services. The findings of the report were extremely encouraging - the company was perceived as excellent overall by it's client base.

The company had also reached the critical point where in order to sustain roles of growth and market penetration, there was a need for the Owner/Manager to delegate more of the day to day operations. It was also discovered that there had been some minor technical problems on past projects that we were unaware of as they had not been brought to our attention in everyday dealings with the clients.

The recommendation of the report, submitted early 1988, was that we implement forthwith a Quality Management System to improve control.

Quality Initiative

By July 1988 we had appointed a consultant to document and assess the company's existing Quality Systems and procedures for compliance with BS 5750 Quality Management Systems, and to provide advice and guidance where improvements where considered necessary.

The initial survey report compiled by the consultant showed many elements of sound practice within the management system adopted by Loy Surveys.
The performance of surveys to client requirements was given specific control by use of the QA form, and having a compact management exercising firm control of the day to day operations within the company.

The existing informal quality systems required to be formally documented to the level of current compliance to ensure that the current good practices were maintained in an orderly manner. The existing controls were then reviewed against BS 5750 and extended as necessary to ensure that all processes operate to the Quality Assurance principles embodied in BS 5750.

It is not my attention to go through all the detailed recommendations and implementation of that initial survey report as these will vary between organisations however I would highlight in particular the need for a systematic approach to training and correct some common misunderstandings that may arise.

The Surveyor being generally a systematic, precise and accurate type of individual has a tendency to confuse accuracy with quality and assume quality will require everything measured to 1mm. Quality management systems demands that you agree and document the accuracy requirements in accordance with the nature of the type of work prior to commencing.

If Quality Assurance can eliminate work by "Getting it right first time" then Quality Control can lead to the creation of work by unnecessary reworking, it is important to learn the difference.

The Control of Change

The continued growth of the firm resulted in the introduction of new Surveyors requiring support and coaching in the way we work.

Everyone has the potential to perform to a high standard but they need just a little help along the way. With this as the objective, the basis for a strong management team was created during 1988 with a commitment to developing it's own full potential through training.

Training

The results of findings into management training reveal that over half of all UK companies appear to make no formal provision for the training of their managers. Considerable resources may have to be devoted to management training. Technical people can become obsessed with detail and slow everyone down around them. Management must help people understand the difference between Efficiency v Effectiveness.

Training should not be haphazard but should be approached systematically. You should identify the training which will be required and make provision to supply this need. To discharge this requirement effectively, you should review all the activities of your organisation which effect quality and ask yourself the question - do the personnel carrying out these activities have the appropriate qualifications, skills and /or experience to fit them for this task ? If the answer is no in respect of any aspect of that question, then you have identified a training need.

The need to maintain adequate records of training, qualThere are a number of circumstances which give rise to training needs: certain tasks require acquired skills, e.g. computing, survey instruments etc. and competence in these activities generally comes from practice. It is therefore essential that the necessary provision be made to ensure that personnel acquire practice under controlled conditions and that their progress towards the established skill level is monitored. The introduction of new equipment or processes necessitates suitable training to familiarise personnel with the new operations. Personnel also need training to equip them to take on additional roles or responsibilities such as communication skills.
ifications and experience of staff is an important element of a quality management system.

Another important area is the need to ensure that all personnel understand the company's quality policy and objectives and are familiar with those aspects of the quality management system which affect their activities.

This brings us to the subject of personal quality which can decide a company's future. It's up to management to inspire each and every employee to deliver a high standard of personal quality. The greatest benefit that people can gain by personal quality is self-esteem -and self esteem is the basis of quality. Management's most important task is to motivate people - the organisation's most valuable resource - to do their best.

Success or the start

On the 4th August 1989 a certificate was issued to certify that the Quality Management Systems of Loy Surveys Ltd had been assessed by Yarsley Quality Assured Firms Ltd and registered as meeting the requirements of ISO 9000/BS 5750 Part 2.

Ref: JGL/QA01/Rev A

This unique accolade has had a positive effect on the moral of staff, without whose efforts this prestigious award would not have been possible.

It also marks the start of a new phase in the development of the company. We can go forward with confidence, but the surveillance visits, required every six months to maintain registration will ensure there is no room for for complacency to set in.

All supplies/services are subject to evaluation and monitoring. The suppliers are made aware of the need to meet and maintain performance standards.

The award of BS 5750 can also be very demanding as it has the effect of increasing the clients awareness and raising his expectations.
The challenge we now face is to live up to the high standards we have set and strive to further improve our services.

The Future

The professional surveyor should not regard the introduction of quality development as a threat to his integrity or judgement, but more as an opportunity to further convince his client to place confidence in his professional ability to meet his needs consistently.

Finally I put forward the case that cheapest is not always best. Better value may be obtained by developing a more committed relationship with a supplier selected for quality potential as much as for price.

"Partnership Sourcing" as a concept is already firmly established in the Japanese economy, and a few British firms have proved the value of partnership sourcing over many years with more moving in this direction.

Conclusion

Our experience proves that the search for excellence is indeed a journey and not a destination. A quality system can be introduced fairly quickly but excellence takes a little longer. We have only started that journey with a system that will measure our progress along the way, may I wish you well on your travels.

APPENDIX A

Some Hallmarks of a Quality Company

* Quality is taken seriously and is part of the company culture.

* Management strives to meet the high standards the programme sets for efficiency and human relations

* There are clearly defined quality goals for all areas. Standards are high. Results are constantly monitored and publicised.

* Quality assurance control is not perceived as a sign of distrust, but rather as a desire to develop and maintain quality.

* Employees are the company's most important resource. The company invests in training and development at all levels in both technical and human skills.

* The company distinguishes between acceptable and unacceptable mistakes. Acceptable mistakes are creative mistakes. They stimulate development, test new knowledge and are a sign of experimentation. Unacceptable mistakes are "sloppy" mistakes. They are unnecessary, expensive and damaging.

* The level of decision making is placed no higher than necessary. Knowledge decisions are made at the level where quality demands can be met.

* Quality evolution takes place not only inside the company but with the customers too.

Ref: JGL/QA01/Rev A

TOPOGRAPHIC SURVEYOR - TRANSITIONING TO THE 90s

Richard B. Marth, Sr.
U.S. Army Engineer Topographic Laboratories (USAETL)
Topographic Developments Laboratory
Surveying Division
Tactical Positioning Branch
Fort Belvoir, Virginia 22060-5546

ABSTRACT

As we transition to the 90s, the tasks of the Army Topographic Surveyor will remain the same - provide control, layout construction, etc. His tools will change. The Automated Integrated Survey Instrument (AISI) will be a total station survey instrument replacing some of the old survey tools. AISI will speed up the tedious task of data logging. Global Positioning System (GPS) will be added to the Army Topographic Surveyors tool box. GPS will not replace existing instruments but will supplement those already used by the surveyor. If not limited by natural surroundings or man-made surroundings, GPS will be used to perform the long legs of control extension. The Army Topographic Battalions have invested their own money in the future by purchasing commercial total stations and GPS receivers. Informal training to support this move in the future has been conducted by USAETL. A training base needs to be established.

INTRODUCTION

Since the earliest days of the United States, Army surveyors have provided support to a growing nation. They helped to expand the reaches of our nation by developing maps, documenting transportation routes, and establishing general survey control and information. This support has been supplied in peacetime and in wartime. Currently the Army is training three types of surveyors - topographic, construction and artillery - who provide survey support. The Topographic surveyor extends national survey control to military areas of operation, establishes 2nd and 3rd order survey control points (SCP) to support construction, signal and artillery units and augments the artillery surveyor. The Topographic surveyor also establishes the initial and closure survey control for navigation systems as well as other survey related services and information required for airfield operation. The Construction surveyor provides construction site positioning, layouts and other relevant information. The Artillery surveyor extends the survey control established by the Topographic surveyor forward to artillery firing positions.

HARDWARE FOR THE ARMY SURVEYOR IN THE 1990s

Hardware for the Army surveyor has been adaptations of civil sector equipment to meet military operational requirements or has been developed to meet specific military requirements. A mix of hardware from these different sources has formed the Army surveyor's toolbox. The new survey equipment to be introduced in the 1990's will augment the current survey tools. In some cases, the new hardware may replace some of the existing hardware.

Current Hardware.
 The surveying tools currently used by the Army surveyor are similar to those tools found in civil surveying crews. Theodolites, Wild T-2s, T-3s and T-16s are the work horses for the Army surveyors. Transits, steel tapes, rods, planning tables and other surveyor's tools fill the Army surveyor's tool box. These basic survey instruments are used extensively by the Construction surveyors.

The Topographic surveyor has some additional tools to support his need to determine positions, distances and azimuths. The Analytical Photogrammetric Positioning System (APPS) allows the Topographic surveyor to determine target coordinates. The APPS use a photographic data base, the Point Positioning Data Base (PPDB) produced by the Defense Mapping Agency (DMA). The Topographic surveyor also has the MX 1502 Satellite Surveyor to establish SCPs. The MX 1502 is a receiver which receives and records TRANSIT satellite broadcast information. The recorded data are sent to DMA for processing with the precise satellite ephemeris. The observation time on station with the MX 1502 is 2 - 4 days and the post-processing by DMA is 4 - 6 weeks. Despite the long start to end time for the survey with the MX 1502, the system has proven to be a good system. The future use of the MX 1502 is limited. The equipment is old and hard to get repaired. The TRANSIT system is also being phased out of operation.

For the tasks of determining distances and azimuths, the Topographic surveyor has a variety of methods and hardware available. The Topographic surveyor has electronic distance measuring equipment (DME). The DMEs are microwave (long range) and infrared (short range) and provide the capability to extend survey control over long distances. In addition to sun and star observations, the Topographic and Artillery surveyors uses the Surveying Instrument Azimuth Gyro Lightweight (SIAGL) to determine azimuth. The SIAGL is a North Seeking gyro system.

The Artillery surveyor also has the Position Azimuth Determining System (PADS) to provide position and azimuth information for the field artillery. The PADS is an inertial system mounted on wheeled vehicles. PADS is initialized at a survey control point established by the Topographic Surveyor. The Artillery surveyor then uses PADS to extend control to firing points. PADS is periodically updated at known survey control points.

The Army surveyors have handheld calculators and Z-248 personal computers for computational support. This equipment is getting old and there are plans to introduce new computer equipment to survey squads and teams.

Automated Integrated Survey Instrument (AISI).
The Automated Integrated Survey Instrument (AISI) is the military equivalent to a commercial total station. The AISI will electronically measure horizontal and vertical angles, and distances. The angular measurement accuracy will be 00.1 second. The measurements will be electronically recorded. There will be two types of AISI, Type I for the Topographic surveyor and Type II for the Construction surveyor. The difference between the two types is the range of the distance measuring capability. The AISI Type I will have a 7 kilometer range for the distance measuring module. The distance measuring module for the AISI Type II will be 2 kilometers. The AISI is currently scheduled for fielding in the 1992 - 1993 time frame.

Computation Support. The Army survey teams and squads will be receiving Army Command and Control System (ACCS) computers. The ACCS computers are Unix based systems. Software designed for the Army surveyor will be written in Ada, the designated DoD software language. The specifications for this software are being prepared.

Global Positioning System (GPS).
The Global Positioning System (GPS) is a DoD satellite based system managed by the Air Force. When GPS is fully operational, there will be 24 satellites (21 operational, 3 spares) in the constellation. The constellation configuration will provide 3 dimensional positioning 24 - hours a day. The satellites broadcast on two wavelengths, L1 and L2, 19 cm and 24.4 cm respectively. L2 use has been primarily for Ionospheric refraction correction. The system information can be degraded to protect the high positional accuracies from immediate enemy use. Thus the positioning accuracies achievable with GPS will depend on the capability of the user's GPS receiver. Standard Positioning Service (degraded satellite information) will be available to everyone and has a horizontal positioning accuracy of 100 meters. Precise Positioning Service (PPS) will be available to authorized users and has 10 meter horizontal positioning accuracy. Current Air Force policy calls for implementation of the capability to degrade satellite information. The survey teams and squads are scheduled to receive GPS receivers.

GPS will meet the stated position accuracy requirements for most of the Army. The Topographic surveyor has the responsibility for establishing geodetic quality control which will require higher order positioning accuracy. The Topographic surveyor will achieve the higher positioning accuracy using GPS absolute positioning techniques and GPS relative surveying techniques. GPS relative surveying techniques can be divided into two groups, carrier phase and code phase. Current civil applications of all of these techniques will need to be adapted to combat situations.

Absolute Positioning. Absolute positioning is a single GPS receiver operation and meter or less positioning accuracies have been achieved. Long station occupation times and post-processing are necessary to improve accuracy. Station occupation times of 4 hours have been recommended as part of the process. The long occupation times give better time averaging of multipath, Ionospheric effects and other effects on the broadcast signal. With PPS, 3 - 5 meter accuracy has been achieved with 4 hour observation times and with 10 minutes of post-processing using the broadcast ephemeris. This accuracy is improved to a meter or less by post-processing with the precise satellite ephemeris. The use of dual frequency GPS receivers can also be beneficial for absolute positioning.

The Topographic surveyor will have military GPS receivers and will be a PPS authorized user. The station occupation time then becomes a key factor. The accuracies achieved with the military GPS receivers will need to be quantified as a function of station occupation time.

Relative Surveying (Carrier Phase). Geodetic quality control can also be established using the GPS carrier phase techniques. These techniques have been used in the civil sector to achieve centimeter level positioning accuracies. The carrier phase techniques require post-processing. Two or more GPS receivers are used for this technique. One receiver is placed on a known SCP and the remote receivers are placed at locations where SCPs are needed. Time tagged information is recorded by all receivers used in the survey and the information is post-processed. In the post-processing, the precise baseline vectors from the known position to each unknown position is determined. This then permits the position of the unknown position to be computed to very high accuracy.

Relative Surveying (Code Phase). For surveys which do not require geodetic quality control, control can also be established using the GPS code phase techniques. These techniques have been used in the civil sector to achieve 3 meter positioning accuracies. The code phase techniques can be applied real-time or used in post-processing. Two or more GPS receivers are used for these techniques. One receiver is placed on a known SCP and the remote receivers are placed at locations where SCPs are needed. Time tagged information is recorded by all receivers used in the survey. At the known position, the difference between the known position and the GPS determined position is used to determine the pseudo range corrections to each satellite. These corrections are then applied to the information recorded at the remote receivers. If a communications link is available, the corrections can be determined in near-real-time as the survey is being conducted and transmitted to the remote receivers.

Although this near-real-time application of GPS appears promising for adaptation to military surveying tasks, questions need to be resolved for the Army surveyor. Among these are the availability of a radio link for corrections, the classification of the broadcast corrections and the maximum separation between known and remote receivers. These questions are being address by the GPS policy makers and are being researched by USAETL.

TRAINING FOR THE ARMY SURVEYOR IN THE 1990s

The use of total station instruments and GPS has and is changing the way commercial surveying is being performed. The introduction of these new instruments into the Army will likewise change the way the Army surveyors conduct their surveying tasks. Training in the use of new instruments always plays an important role in their impact on improving the execution of the tasks they are designed to perform. However training in the operation of the AISI and the use of GPS for surveying is only part of the required training, good training in the basics of surveying will insure the Army surveyors ability to properly execute surveying tasks.

The importance of training in basic principles can not be over stressed when new, automated equipment is introduced. Hi-Tech equipment does not necessarily relate to training of and operation by personnel of lower level ability. For example, a calculator with a simple interest calculation function is very useful. But if the user does not understand the basis for the calculation, the user may not realize that $2000 principle and interest after one year on a $1000 5 percent savings account is wrong. While this example may simply state a potential problem, the complex calculations and the need to determine the quality of a survey are more open to miss interpretation of results. Good training in basic surveying and the interpretation of results will reduce the probability of not discovering errors. Once the basics of surveying are understood, the impact of the new surveying tools, AISI and GPS, should be learned. The new tools will effect training in the areas of survey planning, execution and data analysis. Planning a survey has and will be an important key to efficient, smooth survey execution and an understanding of flexibility offered with the addition of AISI and GPS must be taught. Satellite geometry and the influence of features which reflect or inhibit GPS signals, and other GPS phenomenon needs to be taught as part of the data analysis training.

In anticipation of the fielding of AISI and GPS, Topographic Battalions have purchased total stations and GPS receivers. USAETL has provided informal training. The U.S. Army Engineer School (USAES) is preparing training for operation and use of AISI and GPS for military surveying. The training will be based on the adaptation of civil sector total station and GPS survey techniques to military requirements. USAES is looking for documentation of current total station and GPS surveying methods used in civil sector, as any existing training documents would benefit USAES in the development of total station and GPS surveying courses.

FIELD SURVEY SUPPORT IN THE 1990s

The Army surveyors will continue to support the Army and the nation in peacetime and wartime. These surveyors will be better equipped with the AISI and GPS. The keys to successful completion of required surveys will be their understanding of the basics of surveying and their ability to apply the new instrument to the task. Good training in Army schools and unit reinforcement in both areas will assure that Army surveyors are the best and are fully capable of performing surveying tasks.

TIDAL DATUM COMPUTATION IN LOW TIDE RANGE AREAS

Douglas M. Martin
National Ocean Service
Rockville, Maryland

F. Michael Speed
Carroll I. Thurlow
Conard Blucher Institute
Texas A&M University System
Corpus Christi, Texas

ABSTRACT

Tidal datums are used for nautical charting and for determining the baselines for delineating and/or demarcating state/Federal offshore boundaries and to establish the boundary between state and private ownership of property. Many state and Federal agencies with resource management, jurisdictional, and regulatory responsibilities in coastal areas also rely on tidal datums to accomplish these mandated responsibilities.

National Ocean Service (NOS) procedures for computing tidal datums are based on short series of measurements compared with simultaneous observations at long term (19-year) control stations. NOS control stations are generally located in coastal areas where the tidal signal is predominent in the daily rise and fall of the water.

There are areas along the coasts where the daily water level fluctuations caused by meteorological and oceanographic effects are larger in magnitude than in periodic tidal variations. This situation exists in many shallow water bays, estuaries, and lagoons with relatively narrow entrance channels. The shallow water and narrow channels tend to dampen the tidal signal, reducing the tidal range and amplifying the meteorological and oceanographic effects. Tide stations several miles apart and on the same body of water will not have simultaneously occurring tides, the same number of tides, or the tides will be completely obscured for several continuous days, which makes the selection of a control station for datums computation difficult. This paper describes standard methods of datum computation used by the NOS and alternative methods are compared and evaluated.

OPEN-END ELECTRONIC SURVEY TRAVERSING

Jon E. McReynolds, PLS
Lane County Public Works
3040 North Delta
Eugene, Oregon 97401

ABSTRACT

Unique computer software used to process electronic survey data now make the open-end survey traverse possible. Until recently the closed survey traverse method has been used as a way of detecting some types of survey errors. However, by compiling electronic survey field information with data checking software, errors may now be discovered that were once only certain of being detected by the closed traverse method. One version of an open-end traverse computer program has already been created by the Lane County, Oregon Public Works Department. Whether computer programmers modify existing programs or create their own new programs, software design is now feasible that will enable use of the open-end traverse for many types of surveys.

INTRODUCTION

The author does not contend that the open-end survey traverse be a substitute for the closed traverse, but suggests its use be for local surveys, when closing or networking is not desired. By definition an open-end traverse begins from a station of known or adopted position, but does not end on such a station. Also by definition, and in contrast to the open-end traverse, the ending point of a closed traverse closes on a station with a known relative position (ACSM & ASCE 1978). Despite the differences in their ending points, these two types of traverses have many common elements and their accuracy is dependent upon the same criteria. That criteria, in general, is the survey instructions which are to be followed when executing the work, and the precision with which those instructions are followed (Mitchell 1979). Past use of the open-end traverse has been restricted because necessary calculation checks were limited and not resolute. Also, the open-end traverse could not be adjusted, consequently, surveyors have had little trust in use of this traverse.

Lane County Public Works (LCPW) has developed computer software which will process and check open-end traverse electronic field data for its reliability. Thus, when survey instructions (standards and specifications) are applied to an open-end survey traverse, and the resulting field data is properly processed, the open-end traverse may be used with confidence. The main benefit of the open-end traverse is its cost effectiveness, however there are other unexpected benefits which were produced by the development of this software.

COLLECTING & PROCESSING ELECTRONIC FIELD DATA

For the purpose of this paper, the definition of "electronic data" is that survey field data which is initially collected by a total survey station. The collector, a total station, is a survey instrument that measures distances and angles, and then records them, along with user-entered codes, into an electronic field book, which is called a data collector. The information in the data collector is then downloaded into a computer where appropriate software may produce final measurements, statistics, maps, etc. The following remarks will describe how these processes may be applied to an open-end traverse.

LCPW electronic data collection

The fundamentals of electronic data collection systems are essentially the same, thus only the LCPW electronic data collecting techniques required for processing open-end traverses will be discussed. Those techniques are, simply, measure multiple sets (direct & reverse) of angles at each traverse angle point, and also, measure distances in both directions on all traverse lines. To illustrate the details of these methods, the LCPW data collecting process will be described.

An example from a portion of one of LCPW's traverses is shown below in Figure 1.

Figure 1
LCPW open-end traverse example
(No Scale)

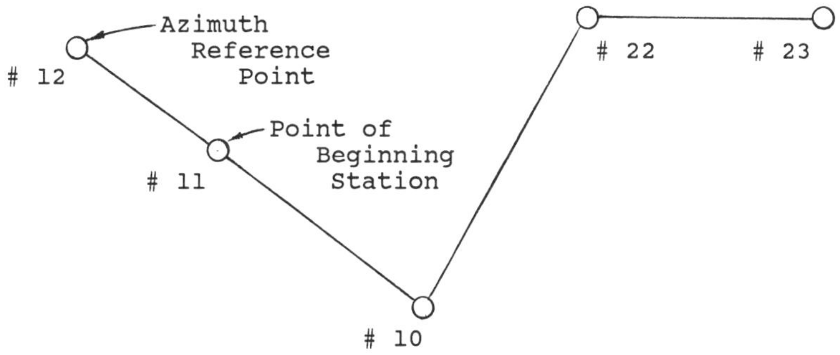

12 — Azimuth Reference Point
11 — Point of Beginning Station
22   # 23
10

As the fieldwork was performed on the traverse in Figure 1, multiple sets of angles were measured at points #11, #10, #22, etc. In addition to measuring angles, distances were measured simultaneously with each theodolite direct pointing. For example, distances were measured from #11 to #10, from #10 to #11, from #10 to #22, etc. Prior to each set of measurements, traverse point numbers and instrument and target heights were also coded and recorded into a data collector. Listed after the coding data are the angles and distances which were also recorded as they were measured. At the conclusion of the fieldwork, the electronic data was downloaded into a VAX mini computer where the data was processed by LCPW-created software. Finally, traverse (and topography) coordinates derived by this process may be transferred to Autocad for final calculations and mapping. In addition to these field and office methods, survey field standards and specifications (see "Standards and Specifications....") were also an essential component required to increase the reliability of the electronic data.

Reliability requirement of LCPW electronic data

LCPW computer software receives two types of electronic data through the field collecting process. The first is "instrument" data, and the second is "user" data. The first type is observed, measured and transferred electronically and is generally free from recording errors and blunders entering into it throughout this entire process. The second type, or user-entered data, is obviously more prone to have human errors introduced into it. If these user codes (which instruct the software data compilation process), are wrong, they may introduce error into the previously error free "instrument" data. Therefore, as the author compiled formulas, calculation procedures, and data processing flow charts for LCPW computer programmers, the goal was to have the software designed to detect user-entered coding errors, thus improving data reliability. After programmers completed the software and it was in use, an obvious bi-product of this design was its ability to check open-end traverses.

Processing LCPW electronic data

There is a significant operation that occurs within LCPW software when it is processing open-end traverse data, which assists in detecting gross errors. That operation occurs in the first stages of processing, when electronic data is reduced to one final angle and distance for each traverse point and ensuing course. This phase of the software has a data "linking and overlapping" procedure which insures that traverse point sequencing and subsequent data calculation results will be reliable.

Point linking. LCPW software began processing of the example traverse (Figure 1) data with the designated traverse sequence #12 - #11 - #10, and noted the occupied instrument point (#11) and foresight point (10). The program then designated these points as the backsight point number and the instrument point numbers, respectively, for the next traverse point sequence. Hence, for the next sequence, the software searched for and located points #11 - #10 - #22, which satisfied the "linking" requirement. The point linking results, from the traverse represented in Figure 1, are displayed in the horizontal statistical report in Figure 2. This Figure first displays the traverse point number sequence, then horizontal angle measurement statistics for that sequence are shown immediately after. These statistics are derived from the three numbered sets of ANGLES, and are: the MEAN angle, the angular DIFFERENCE FROM that MEAN and each of the three sets of angles, and the STANDARD DEVIATION.

In addition to angular precision, the procedure that processes the data listed in Figure 2 also readies that data to be tested in the COGO segment of the software for two types of field "point" coding blunders. For example, when field personnel leave a gap in the traverse point coding sequence, the COGO portion of the LCPW software stops processing and indicates a sequence error. The other blunder, an overlap in the traverse point sequence, is also exposed in the COGO phase when that software indicates that a point has a value that is being changed to a new value.

Figure 2
LCPW ELECTRONIC DATA COLLECTION STATISTICAL REPORT
HORIZONTAL ANGLES

```
POINTS =        #12           #11            #10
  MEAN =    183° 41' 15.6"   STD. DEVIATION = ± 0° 0' 0.3"
  SET         ANGLE          DIFF. FROM MEAN
   1       183° 41' 15.1"      -0° 0' 0.5"
   2       183° 41' 15.5"      -0° 0' 0.1"
   3       183° 41' 16.2"       0° 0' 0.6"

POINTS =        #11           #10            #22
  MEAN =     90° 06' 57.3"   STD. DEVIATION = ± 0° 0' 0.4"
  SET         ANGLE          DIFF. FROM MEAN
   1        90° 06' 57.6"       0° 0' 0.3"
   2        90° 06' 56.4"      -0° 0' 0.9"
   3        90° 06' 57.8"       0° 0' 0.5"
```

Overlapping traverse courses. Mark to mark (slope) distance results, from one course of the traverse represented in Figure 1, are shown in Figure 3.

Figure 3
LCPW ELECTRONIC DATA COLLECTION STATISTICAL REPORT
MARK TO MARK DISTANCES (IN FEET)

```
          POINTS =         #11           #10
       MEAN = 2653.383   STD. DEVIATION = ± 0.002
       DIRECTION      DISTANCE     DIFF. FROM MEAN
       #11 -> #10      2653.39          0.007
       #11 -> #10      2653.39          0.007
       #11 -> #10      2653.38         -0.003
       #10 -> #11      2653.38         -0.003
       #10 -> #11      2653.38         -0.003
       #10 -> #11      2653.38         -0.003
```

The two POINT numbers listed first in Figure 3 represent the traverse course that was measured in the field. The heading DIRECTION indicates the direction that the course was measured in the field. For example, in Figure 3, the first three distance measurements, from point #11 to point #10, were obtained when point sequence #12 - #11 - #10 was measured; and the second three measurements, from point #10 to point #11, were obtained when point sequence #11 - #10 - #22 was measured. Thus, the course between #11 and #10 becomes an "overlapping" course between the adjacent traverse point sequences. Also in Figure 3 are the MEAN distance and the STANDARD DEVIATION, and the DIFFERENCE FROM the MEAN for each of the six displayed DISTANCES.

Statistics in Figure 3 will also expose two types of field errors, as well as indicate distance precision. The first type of error is revealed when the software attempts to overlap point sequences as discussed above. For example, if the Figure 1 traverse sequence #11 - #10 - #22 had erroneously been coded #12 - #10 - #22, the statistical report would list DISTANCE #10 -> #12 separately from DISTANCE #11 -> #10. Also, instead of listing six distances for each course, there would only be three in each listing, which would be three distances less than required. This type of coding error is exposed only in this statistical report.

The second type of error exposed by this overlapping process is a field logistic blunder. For example, if an attempt was made to measure traverse point sequence #11 - #10 - #22 in the field, but mistakenly a field point other than #10 was occupied with the total station, or the target was mistakenly placed on a field point other than #11, then two different distances (DISTANCE #11 -> #10 and an erroneous DISTANCE #10 -> #11) would be averaged together in the distance report. Therefore, an error message would be displayed because the DIFFERENCES FROM the MEAN, and the STANDARD DEVIATION, for traverse course #11 -> #10 would be grossly over the maximum limits.

Standards and specifications for traversing

Standards and specifications are a necessary component of any electronic data collection process, whether for an open-end traverse or a closed traverse. This occurs because data quality, electronic or otherwise, is influenced by the specifications used when collecting it in the field, or when processing it in the office. Traverse specifications consist of four general components, which are: 1) traverse geometry; 2) instrumentation & calibration; 3) field procedures; and 4) office procedures.

Accuracy verses guesswork. As discussed prior, the instrument element of electronic data collecting is generally error free. This tends to overshadow all other specification components, when it actually satisfies only part of the "field procedures" component. However, not any of the four areas of specifications may be disregarded, and if any of these areas are breached, the accuracy of an entire survey may be reduced to the level of the weakest specification component. To determine which level of accuracy the electronic data may qualify for, it must first be determined which level of specifications were used when obtaining the data (this would be true for determining the standard of a very low order survey as well for a high order survey). Hence, if any of the four specification components become unknown, the accuracy of the survey becomes guesswork.

Error of closure. The error of closure of a survey is generally used as a tool (office procedures specification) for detecting blunders. After blunders are found and removed, the remaining errors are generally treated as systematic errors, which are reduced through an error of closure adjustment. However, if a certain specification level is followed, systematic errors can be detected and may nearly be eliminated when performing a survey.

The open-end traverse concept was tested by the author over a four year period, while performing closed traverse surveys with a Wild T2000 theodolite and DI5 edm. Certain traverse specifications were decided upon and used for the work performed, and open-end traverse results were checked by closing them. However, traverse calculations were first performed as open-end calculations and blunders were detected and eliminated as described prior. Finally, after the calculations were completed, the error of closure was calculated and the open-end traverse results checked. In cases where the use of the open-end traverse was applicable (i.e., road construction surveys, topography surveys, road right of way surveys), the results were always adequate without adjusting for the error of closure. As suggested in the text <u>Surveying</u>, "...for certain kinds of surveys, particularly those of low precision, it is unnecessary and impracticable even approximately to eliminate errors" (Davis, Foote, & Kelly 1966).

Benefits of LCPW electronic data processing software

The obvious benefit of LCPW electronic data processing software is its ability to check an open-end survey traverse. The open-end traverse may be desirable for some types of traversing, and is also cost effective when compared to surveying extra courses that may be required to close a traverse. Consequently, in some situations the open-end traverse may be very attractive from a financial viewpoint.

In the process of creating the LCPW software, other secondary benefits were also realized. For example, because of software logistics, survey field personnel are allowed great flexibility in their field procedures. Since data is linked together in the computer program, field operations do not have to be linked, that is, points do not need to be surveyed in sequence. Thus, in field operations, occupation of random points along a traverse is allowed as needed when considering topography, weather, and other uncooperative circumstances.

It was also found that error free (as defined in "Reliability requirement...") radial side ties can also be made with the same software functions that were used for error free traversing. For example, if radial side ties are measured when occupying and back sighting along a main traverse, the software compares that backsight distance to the main traverse course distance, thus confirming that the measured course and the coded course are one and the same. Hence, users can be assured that the field backsight and instrument occupation points are correct, therefore assuring that the ensuing side tie azimuths will also be correct.

There were many other benefits realized from the design and use of the LCPW software, which are too numerous to mention here. Many of them relate to cost effective field surveying that was derived from the flexible software design.

CONCLUSION

LCPW software will inspect incoming open-end traverse electronic data through its "linking and overlapping" data compilation process. Through this process it checks the precision of traverse measurements and the validity of user-entered codes. These measurement and code checking procedures, along with use of correct survey Standards and Specifications, allow the often cost effective open-end traverse to be used with confidence.

ACKNOWLEDGEMENTS

The author wishes to acknowledge the valuable suggestions given by Gary Bricher, Lane County Public Works Computer Programmer, who was also the designer of the LCPW electronic data compilation program.

REFERENCES

ACSM & ASCE: American Congress on Surveying and Mapping and the American Society of Civil Engineers 1978, Reprinted 1981, <u>Definitions Of Surveying And Associated Terms</u>, P. 172

Mitchell, H.C. 1948, Reprinted 1979, <u>Definitions of Terms used in Geodetic and other Surveys</u>, Special Publication No. 242, P. 2

Davis, R.E; Foote, F.S; Kelly, J.W. 1966, 5th ed, <u>Surveying</u>, McGraw - Hill, New York

A Formal Model for Land Titles and Interests in Land

Ricardo Javier Moreno
University of Maine
Orono, Maine 04469-0110
(207) 581-2187
RICARDOM @ MECAN1

Abstract

Massive quantities of land information, a large percent being redundant, are created everyday with no formal linkage mechanism. Decisions for determining land use, controlling land records, and implementing mapping projects may be achieved more efficiently with a legally supportable cadastral information system (LSCIS). An LSCIS is envisioned as having two interconnected primary components; a software system for efficiently integrating a wide range of measurement observations made to locate boundary corners and a software system for managing information about individual cadastral parcels. Users of land information systems requiring accurate boundary and ownership information might find an LSCIS as the most appropriate and efficient means for providing the cadastral components of the land information system.

This presentation will describe a formal approach for organizing information concerning land titles and interests in land. Research is being pursued to develop specifications for the relationships among land recordation, types of land conveyances, mortgages, leases, sublets, wills, and similar processes and instruments. A structured approach is being followed to facilitate the programming of such information in an LSCIS.

STABILITY ANALYSIS OF STRUCTURES OVER AND AROUND THE EGYPTIAN SUBWAY

Osama M. Moussa
Assistant Professor
Military Technical College
Kobry El-Kobba, Cairo, Egypt

Mohamed M.H. Attabi
Assistant Professor
Faculty of Engineering
Ain Shams University
Abasia, Cairo, Egypt

Ehab H. Soliman
M.Sc. Student
Military Survey Department
Kobry El-Kobba, Cairo, Egypt

ABSTRACT

Precise levelling were carried out along a part of the main road of the subway's surface. Each of these levelling started from one fixed point. These procedures were done periodically to check the difference between each set of elevations. The results were analyzed to determine the stability of the main road over the Egyptian subway. On the other hand, different trignometric levelling were established to obtain the effect of the underground structure on the surrounding buildings.

INTRODUCTION

Due to the rapidly increase of the population in Egypt, especially the Capital (Cairo), the ministry of transportation and SOFRETU Company (French Transportation Consultant Company) had been studied the traffic problems since 1964 and they concluded to establish three main subways. The first subway, provincial subway, was established in 6 years period, starting at the beginning of the year 1982 and ended at the end of the year 1987. The main reasons of constructing such subway are to decrease the traffic volume and capacity inside the capital as well as to minimize the air pollution especially during the rush hours. The length of this subway is about 43 Km long and it extends from Helwan, South of Cairo, to El-Marg, north of Cairo. It contains thirty three metro stations, five of them are underneath the earth's surface.

Two set of precise levelling observations were analyzed and the standard deviation for each levelling station was determined. The amount of the error for the best estimator of each elevation, levelling station, as well as the amount of error for the trignometric levelling were found to be satisfactory.

STATISTICAL PROCEDURE

The method of least square adjustment was carried out to determine the stability of structures over and around the Egyptian subway. Two groups of the military survey department established two precise longitudinal levelling along a part of the main road of the subway's surface. Each one of these levelling started from one fixed point, bench mark which was established by the civilian survey department.

The elevation of each levelling station was determined periodically every month after the construction of the surface's road. The precise levelling were carried out in a period of 6 months and they were done twice and three times during December 1987 and January 1988, respectively.

The best estimator, $\hat{X}$, for each elevation can be determined by using the following formula:

$$\hat{X} = \overline{X} = \frac{1}{n} \sum_{i=1}^{n} x_i$$

where $\overline{X}$ is the mean or the average value for different elevations x_i and n is the total number of repeated measurements for x. On the other hand, the standard deviation (standard error), $\sigma_{\overline{X}}$ (the best measure for the amount of error in $\hat{X}$) can be computed by using the following formula:

$$\sigma_{\overline{X}} = (\frac{\sum_{i=1}^{n} v_i^2}{n(n-1)})^{1/2}$$

where v_i is called the residual errors of the measurements x_i and is defined by:

$$v_i = \overline{X} - x_i$$

The standard deviation is sometimes called "root mean square error" and usually abbreviated by "RMS".

RESULTS AND ANALYSIS

The resulted values of the mean and the standard deviation for each elevation for the two set of precise longitudinal levlling were tabulated in Tables 1 and 2.

Table 1: Mean and standard deviation for different elevations of the first precise longitudinal levelling

Station	Mean (in m) $\overline{X}$	R.M.S. (in mm) $\sigma_{\overline{X}}$
A	20.30449	0.44
B	19.61232	0.34
C	19.73104	0.27
D	19.55192	0.39
E	20.24681	0.25
F	20.98197	0.20
G	20.51698	0.26
H	21.44452	0.76
I	21.25426	3.09

Table 2: Mean and standard deviation for different elevations of the second precise longitudinal levelling

Station	Mean (in m) $\bar{X}$	R.M.S. (in mm) $\sigma_{\bar{X}}$
1	22.51500	0.32
2	24.21795	0.02
3	19.63482	0.21
4	22.27841	0.36
5	19.50400	0.38
6	19.75878	0.27
7	19.95968	0.48
8	19.91450	0.34

The degree of precision, degree of reliability of $\hat{X}$, for each elevation was found to be satisfactory, less than 3.2 mm. Most of R.M.S. values were found to be less than 0.77 mm except the value for station I. Different trignometric levelling were carried out for different buildings around the Egyptian subway. The difference in elevation for about fifteen building structures was found in the range of 0.11 mm to 0.15 mm. From the previous analysis, the main road over the Egyptian subway as well as the surrounding buildings were found to be stable.

DYNAMIC POSITIONING SYSTEMS PHOTOGRAMMETRIC TEST COURSE
Anthony Niles
U.S. Army Engineer Topographic Laboratories
Fort Belvoir, Virginia 22060-5546

ABSTRACT

Many electronic dynamic positioning systems that use a variety of technologies, such as laser ranging, UHF signal ranging and satellite ranging, are currently available. With the advances in technology, these systems are undoubtedly becoming more and more accurate. In fact, real-time submeter accuracy using the differential Global Positioning System (GPS) in a boat, aircraft, or ground vehicle is not unrealistic. However, verifying positional accuracy at this level of accuracy of a moving platform is difficult. Close-range terrestrial photogrammetry offers a proven and accepted method for this verification, using mostly off-the-shelf components and technology. Such a system is being developed at the U.S. Army Engineer Topographic Laboratories.

INTRODUCTION

The U.S. Army Corps of Engineers (USACE) uses a variety of electronic positioning systems, primarily in hydrographic surveying, with accuracies ranging from one to three meters. However, GPS systems of decimeter accuracy or better are currently being developed, and it is anticipated that these systems will be widely used within USACE. The better accuracies are generally accepted within USACE simply through repeated use and comparison with other systems. Actually determining the accuracy of time-tagged positions to a few meters or less is difficult; the faster the system is moving and the higher the test accuracy, the more difficult verification becomes. The U.S. Army Engineer Topographic Laboratories (USAETL) thus began searching for a reliable method to test the accuracies of dynamic positioning systems.

Such verification of positioning systems would require comparison with a system of known and accepted accuracy, or a "truth" system. This system must be capable of approximately two centimeters accuracy while moving at 10 to 15 miles per hour. This reflects the anticipated accuracy of real-time carrier phase GPS and typical speeds of survey boats. Due to limited resources and manpower, the system also must have a total cost of no more than $100,000 and must be able to be operated by non-experts.

USAETL considered several methods, such as laser ranging and comparison with an inertial system. However, photogrammetry was eventually designated as the method that would provide the needed accuracies and could be obtained within the practical constraints. In a study performed by Dr. Kam Wong of the University of Illinois, a positional accuracy of two centimeters was deemed possible using two fixed cameras viewing an area approximately 400 by 400 feet. The speed of the test vehicle could be up to 10 miles per hour. Dr. Wong was instructed to investigate the use of charge-coupled device (CCD) array cameras, since such a digital system would eliminate film processing and provide near-real-time positions. However, an off-the-shelf CCD array system was not available and some custom software develpment was needed. Thus the cost and simplicity of use constraints were not met, and it was decided to use a conventional filmed-based system.

Figure 1 on the following page illustrates the concept of the photogrammetric test system. Two cameras, mounted on tripods, view a designated test area stereoscopically. A vehicle with the positioning system to be tested passes through the test area while logging test positions. A radio link synchronizes the positioning system outputs with the camera shutters, thus producing photographs at the instant that the positioning system logs coordinates. The true coordinates are plotted from the photos and are then compared to the test positions.

SYSTEM DEVELOPMENT AND LOCATION

Shortly after the study performed by Dr. Wong, the Tennessee Valley Authority (TVA), based on their extensive experience with close-range photogrammetry, was tasked to develop the photgrammetric test system. The project was divided into five general tasks:

1. Select and procure the cameras.
2. Develop the test system procedure.
3. Perform tests at home site with existing equipment.
4. Implement system at test course.
5. Perform tests with system at test course and instruct USAETL personnel on use of the system.

A suitable location for the test course was found on the grounds of the National Institute of Standards and Technology (NIST) in Gaithersburg, Maryland. The area has clear unobstructed views for the cameras and has favorable terrain relief needed for precise determination of three-dimensional coordinates. Also, personnel at NIST are available for validation of camera and photogrammetric target monuments.

TEST COURSE

The test course, as shown in Figure 2, is an area 400 by 450 feet. The two cameras are positioned 400 feet apart on the front line. Eight photogrammetric targets are positioned along two parallel lines within view of the cameras. The vehicle with the test positioning system travels between the two lines of targets, logging test coordinates and simultaneously activating the camera shutters through a radio link. Each target consists of a four foot vertical range pole with two small spheres, one at the top and one near ground level. This design will be easily identifiable in a photograph and will give unambiguous points for stereo plotting. The layout of the targets and the test course produces the optimum geometry for obtaining the needed accuracies from plotting.

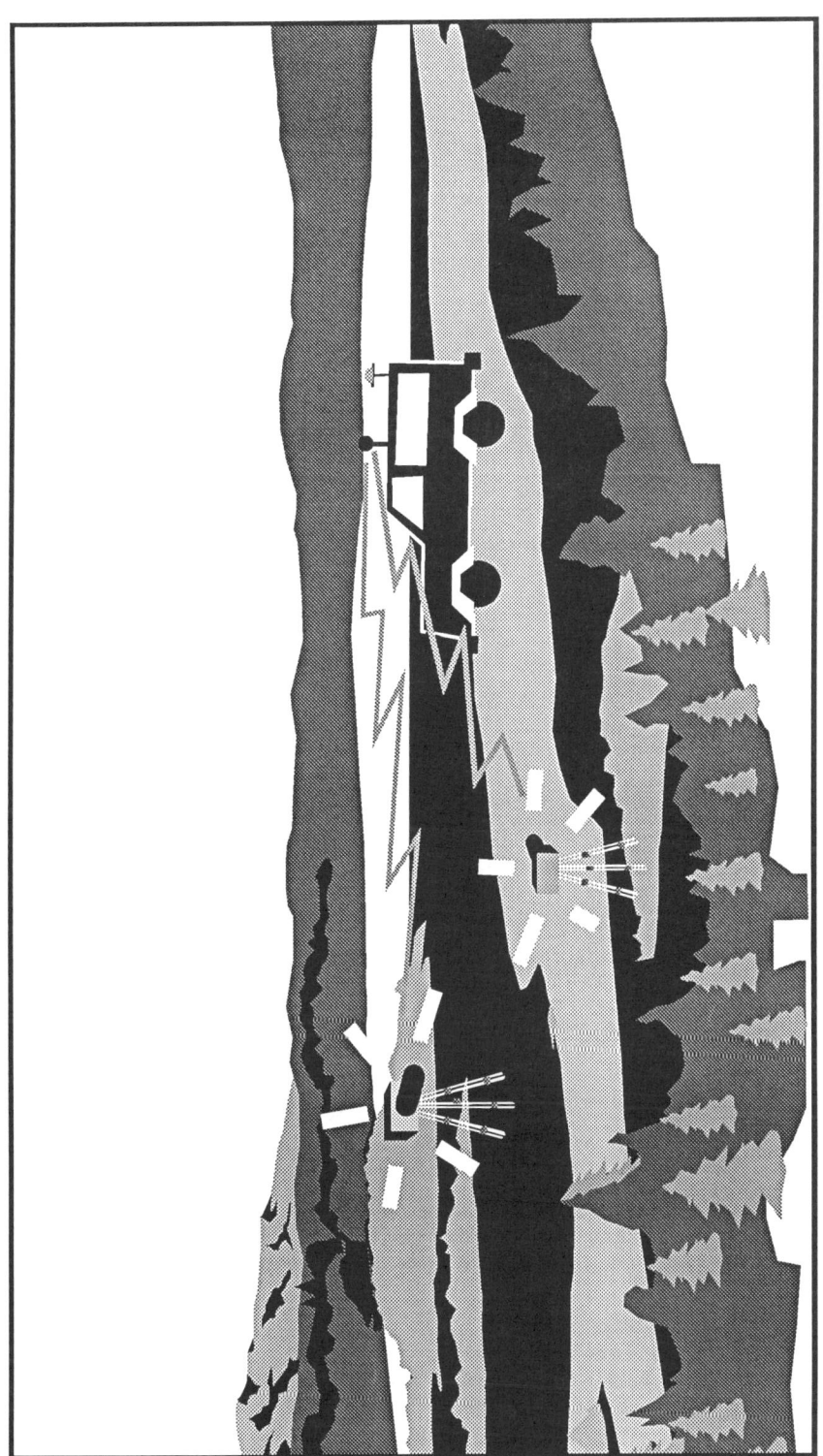

Figure 1 Photogrammetric Test System Concept

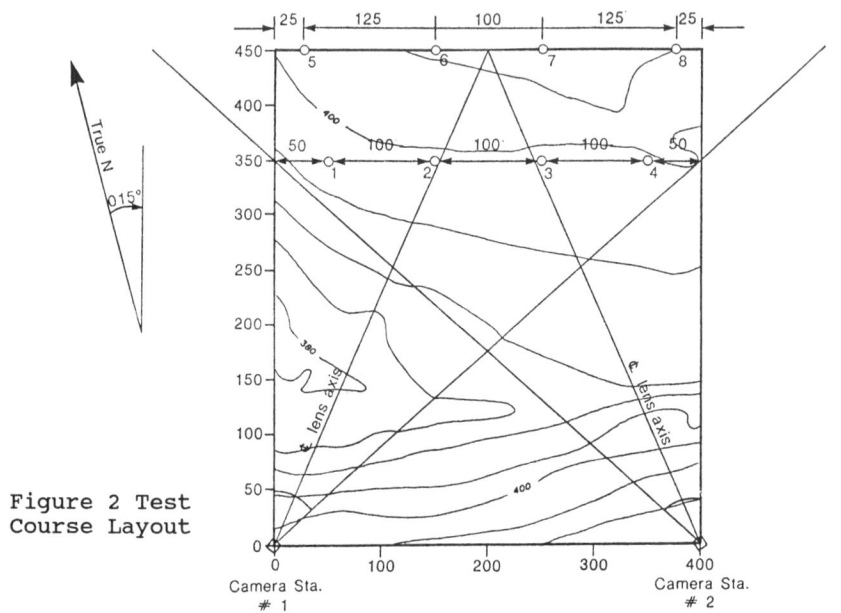

Figure 2 Test Course Layout

TIME-CORRELATED PHOTOGRAMMETRIC SYSTEM

Most positioning systems output coordinates at some fixed time interval. This signal also will be used to activate the camera shutters through an on-board radio transmitter. The cameras to be used for this project are Rolliemetric 6006 metric cameras with built-in grid, focusing stops and factory calibration. The initial positioning system to be tested will be a GPS unit since USAETL is currently developing such systems for hydrographic surveying. The GPS units to be used will have a pulse-per-second output which will be amplified and then transmitted to the camera driver. It is anticipated that other, non-satellite based systems, such as UHF ranging systems, also will be tested, and these also have output pulses that can be used to drive the cameras.

Time Correlator

In order to achieve the precise synchronization needed for the 2-5 centimeter positioning accuracy, timing between the positioning system and the cameras of less than a millisecond is needed. This requires that the transmitting and mechanical delay due to the shutter be determined. This delay is then used to advance the transmitted signal to the cameras.

A time correlator was developed by Dr. K.S. Yang of the University of Illinois to measure this delay. This device has a panel of six rows of high-intensity LEDs. Upon receiving the start signal, a set of counters begins accumulating pulses at a precise rate of 5000 Hz. Thus each pulse represents 0.2 milliseconds of time. The contents of those counters are displayed on the top three rows of LEDs. The rows are arranged in 1X, 10X and 100X unit intervals, each unit being exactly 0.2 milliseconds. The start pulse also is used to activate the camera shutters. A photograph of the correlator panel is then taken. After the film is developed, the exact time the shutter opens, as measured in unit intervals, is indicated by those LEDs that are lit. The bottom three rows of LEDs are used to measure the time the shutter stays open. The LEDs in these rows are lit individually in sequence at the unit interval rate. Each LED stays on for precisely 0.2 milliseconds. The first LED in the fourth row lights up in synchronization with the start pulse. The total number of LEDs seen lit within this group on the developed film is the aperture time measured in terms of the unit interval.

System Operation

A clock within the transmitter is then synchronized with the output pulses of the positioning system. This is done by observing the signals on an oscilloscope and setting the transmitter clock accordingly. The time delay determined by the time correlator is then entered into the transmitter. The camera pulse is then advanced according to the transmission and shutter delay, producing a photograph at the instant of the positioning system output.

SYSTEM DEVELOPMENT STATUS

The complete photogrammetric test system was to have been developed by September 1990. However, some errors were found in the custom options in the cameras shortly after the cameras were received from the manufacturer. TVA is awaiting receipt of the corrected cameras, at which time in-house tests with the time correlator and transmitting unit will begin. It is anticipated that the system will be installed on the test course and trial runs will begin in April 1991.

As mentioned previously, the initial system to be tested will be a differential GPS system. The transmitter and time correlator will initially be configured to receive signals from an Ashtech 12-channel receiver. Tests on other terrestrial electronic positioning systems that are used for hydrographic survey also are anticipated. Since these systems commonly involve range transmitters located several miles away, the photgrammetric system may be ported to a harbor area in which the cameras are mounted aboard a survey vessel and the photgrammetric targets are mounted on shore. The cameras thus become mobile units moving with the positioning system to be tested.

REFERENCES

Wong, Kam W. 1989, <u>Precise Dynamic Photogrammetric Positioning Systems</u>, Contract Number DAAL03-86-D-0001 with the Army Research Office.

Yang, K.S. 1989, <u>An EPS/GPS Time Correlator</u>, Contract Number TV-79383T with the Tennessee Valley Authority.

EVIDENTIARY ADMISSIBILITY AND RELIABILITY OF PRODUCTS GENERATED FROM GIS

Harlan J. Onsrud
Department of Surveying Engineering
University of Maine
Orono, Maine 00469

ABSTRACT

In modern cartographic and surveying practice, maps and plats traditionally maintained as paper documents are now being maintained in digital form in computerized files. Potential errors associated with geographic information systems and other computerized spatial handling systems include input, hardware, software, and perception errors. Because of these potential errors and because digital data is relatively easy to alter with little or no trace, the legal reliability and stability of data from computerized land information systems have become important issues for users and potential users of such systems. General rules for the admissibility of computer generated evidence in court are discussed within a framework of relevancy, authentication, and hearsay considerations.

INTRODUCTION

Developers and managers of automated land information systems are usually more concerned with hardware, software, database, and operational issues than they are with the legal, institutional and social aspects of using such systems or the products generated from them. One among several important social considerations often overlooked in the rush to invest in automated geographic data processing systems is whether records contained within such systems or products generated from them might be needed as evidence in court.

Geographic information systems are presumably being used to increase the quality and efficiency of decision making processes. For instance, a land information system operated by a local government might be used to aid in making decisions regarding tax assessments, zoning and districting delineations, permitting, distribution of social program services among the population, when and where to schedule maintenance of aging infrastructures, siting of new utilities, routing of school buses and emergency vehicles (police, fire, ambulance), and numerous similar decisions. Many of the decisions made using data from a public GIS may affect the daily lives and well-being of a community's citizens. Therefore, adversely affected citizens may chose at times to challenge the reasonableness of those decisions. Although intended to increase the quality of decisionmaking, the "newness" of these innovative systems and our lack of institutional experience with them make them potentially subject to misuse and abuse.

Claims may be made by a challenger that the accuracy or truth of the underlying data in the GIS upon which a decision has been based is highly questionable (e.g. the precision of the data is highly inappropriate for the decision which was based on that data, significant blunders exist in the data, lack of system security, etc.). Claims might be made also that manipulation or processing of the data has been carried out in a manner designed to arrive at results desired by governmental officials rather than carried out in an unbiased manner (e.g. selective choice of a search or buffer radius was used to establish a biased pool of comparable properties for tax assessment or zoning purposes, selective generalization or aggregation of data was used until a configuration supporting a desired result was achieved, computerized "gerrymandering", etc.). Additionally, the electronic data processing may have resulted in input, hardware, software, and perception errors of varying significance.

If the data contained within or the products generated from a specific operational GIS eventually are shown to be readily and successfully challengeable in court, much of the investment in the system may be lost. Decisions based on data from the system generally will no longer be considered reliable by the public. Therefore, land information system developers and managers should be aware of the general requirements for admissibility of computer generated evidence in court as well as be prepared to convince a panel of jurors of the reliability of the system.

Private firms and utilities which use automated mapping or geographic data processing systems to aid the productivity and decision making of their firms need to be just as aware of admissibility and reliability concerns. Increased liability exposure and the need to admit computer generated products into evidence to defend against possible tort actions (e.g. negligence and breach of contract claims) may provide added reason for these firms to be aware of such issues.

ADMISSIBILITY OF EVIDENCE

Evidence is the material offered in court to persuade the trier of fact about the truth or falsity of a disputed fact. Today, all federal courts adhere to the rules for admissibility published in an administrative law volume titled the *Federal Rules of Evidence*.[1] State court systems adhere to similar sets of published rules. The *Federal Rules of Evidence* will be used for illustrative purposes in this discussion because many state court systems have rules of admissibility very similar to the federal rules and a certain proportion of cases involving the admissibility of computer files would be heard in the federal courts

Relevancy of the Evidence ·

The question of admissibility of evidence arises when one party offers evidence into trial and the opposing party objects. The first criteria in overruling the objection is that the evidence must be relevant to a proposition in issue. If the evidence has "... any tendency to make the existence of any fact that is of consequence to the determination of the action more probable or less probable than it would be without the evidence..." it is relevant and is admissible unless excluded by a specific rule. [2]

Hearsay Rule

The most frequently attempted method of excluding computer-generated exhibits, even though the exhibits may be relevant to an issue in dispute, is through the "hearsay rule". Hearsay is an "... oral or written assertion... other than one made by the declarant while testifying at the trial or hearing, offered in evidence to prove the truth of the matter asserted." [3] The hearsay rule states that hearsay is inadmissible unless the evidence qualifies under a hearsay evidence exception. [4]

Data files stored in a computer and printouts generated from those files are almost always hearsay. Printouts are seldom used for other than proving "the truth of the matter asserted." In addition, unless an individual happened to have designed and manufactured the computer hardware, wrote the GIS software, and carried out the product generation or database manipulation procedures involved in the dispute, the computer generated printout (i.e. the written assertion by the individual declaring its truth) will almost always be deemed hearsay. Therefore, to be admitted into evidence, computer printouts must almost always qualify under one of the hearsay exceptions.

a. Business Records Exception

The hearsay exception most often applicable in successfully admitting computer-generated printouts into evidence is known as the business records exception. *Federal Rules of Evidence* 803(6) states:

> The following (is) not excluded by the hearsay rule, even though the declarant is available as a witness:
> **(6) Records of Regularly Conducted Activity.** A memorandum, report, record, or data compilation, in any form, of acts, events, conditions, opinions, or diagnoses, made at or near the time by, or from information transmitted by, a person with knowledge, if kept in the course of a regularly conducted business activity, and if it was the regular practice of that business activity to make the memorandum, report, record, or data compilation, all as shown by the testimony of the custodian or other qualified witness, unless the source of information or the method or circumstances of preparation indicate lack of trustworthiness...

Although computer printouts on paper made at the time the relevant electronic data was collected or processed might be more convincing to a judge or jury, the rule does not require this. A data compilation in any form is acceptable. Therefore, if certain kinds of digital files are always archived in the course of doing business, it is sufficient that the digital data was recorded at or near the time in question. If the date of electronic recording of the specific data can be reliably established, the computer-generated printouts of that data on paper for presentation in court may be made at any time. [5]

To be relevant and admissible, computer files created in the regular course of business must necessarily be authentic. Authentication "... is satisfied by evidence sufficient to support a finding that the matter in question is what its proponent claims".[6] An illustration of authentication evidence for computer records conforming with the requirements of this rule is "...evidence describing a process or system used to produce a result and showing that the process or system produces an accurate result.".[7] This illustration has generally been understood by the federal courts to require that "[t]he proponent of the evidence must authenticate a computer generated business record by showing the input procedures used to supply information to the computer, the tests that were used to assure the accuracy and reliability of both the computer operations and the information that was supplied to the computer, and the fact that the computer record was generated and relied upon in the regular course of business. [8]

Thus, the trend by the courts has been to set additional foundational requirements which must be met before computer-stored business records are authenticated. The authenticating witness does not have to be one of the programmers involved in developing the software. However, the authenticating witness must be familiar with all phases of the field and office procedures which produced the product or result in contention and be able to explain succinctly why it is that errors and mistakes are unlikely to have crept into the system. These errors include "...errors in perception (e.g. misreading or misinterpreting data fed into the system), errors in input, errors associated with inadequate hardware security, errors caused by hardware, and errors associated with computer software."[9] Therefore, GIS managers should be prepared to provide testimony that errors in each of these areas are highly unlikely for their computer-generated evidence in the factual situation before the court. This may be problematic.

The authentication requirements outlined are rather stringent and place a substantial burden on anyone trying to admit computer records into evidence. As a result, "(s)ome courts have ignored suggestions calling for the creation of special foundational rules for authenticating computerized evidence. Instead, these courts require only a custodian of records to testify that the computerized records were kept in the regular course of business."[10] In such courts the custodian need only testify that computer-generated records are what they purport to be and need not attest that the records are accurate. The judgement on the degree of accuracy and reliability to be accorded to the evidence is left almost entirely to the trier of fact. This judicial approach appears to run counter to Rule 803 (6) cited previously which negates the admissibility of evidence if "the source of information or the methods or circumstances of preparation indicate lack of trustworthiness."

To avoid the harsh authentication tests, other judges have moved towards allowing the admission of computer-generated evidence under some circumstances through judicial notice. [11] This is likely to occur only with computer-generated printouts from off-the-shelf computer programs which have been widely used throughout business for an extended period of time. Judges have reasoned in such instances that if a software system wasn't reliable, it would not have lasted in the marketplace. However, it would be unwise for GIS managers to assume that judges might take judicial notice of the reliability of complex commercial GIS software in its current states of evolving development.

b. Other Hearsay Exceptions

There are 23 specific exceptions to the hearsay rule as well as a general exception provision. [15] Of the specific hearsay exceptions, the one likely to find the next greatest utility in the admissibility of GIS evidence is the "public records and reports exception."

Numerous public agencies at the federal, state, and local levels are establishing automated land information systems for a variety of purposes. When a printout submitted for admission is a copy of electronic files collected and maintained by a public agency as a public record or document (i.e. "... a data compilation, in any form, of public offices or agencies, setting forth .... the activities of the office or agency, ...") and the printout is certified as a correct copy by a custodian of the records or some other authorized person, the printout is self-authenticating and no extrinsic evidence of authenticity is necessary. [13] For those local governments claiming proprietary interests in their GIS digital files and the products generated from them, this exception is not likely available since the records are not maintained as "public records".

If all else fails and a computer-generated record is inadmissible due to failure to meet a hearsay exception or failure to meet authentication requirements, the printout may be admitted in some jurisdictions for limited purposes if relied upon by an expert witness. The data or printout is then admitted only for the limited purpose of showing the basis for the expert's opinion and not for the purpose of showing the truth of the data on the printout. [14]

CONCLUSIONS

From a pragmatic legal perspective for a private firm or local government, digital files of spatial data and the products generated from them differ from maps and plats traditionally maintained as paper documents in several respects. The hearsay rule is almost always applied to electronic data files and the products generated from them. As a result, a hearsay exception such as the business records exception or the public records exception must generally be met for GIS generated products to be admissible in court. Authentication as a condition precedent to admissibility also tends to be more complex and difficult for computer printouts. Finally, the reliability and believability of computer printouts may be more difficult to convey to a jury and the general public. To ensure that their products and services conform with the constraints imposed by the legal system, GIS software developers and system managers need to become better aware of these constraints and accommodate them.

ACKNOWLEDGEMENTS

Significant portions of this article were drawn directly from Onsrud, H.J. and R.J. Hintz, "Evidentiary Admissibility and Reliability of Automated Field Recorder Data" (1991) published in *Surveying and Land Information Systems*. I wish to thank Barry Blanchard and Po-Siu (Paul) Hsu for their aid in collecting background articles for this work.

NOTES

1. Federal Rules of Evidence, West Publishing Company, 1990.
2. Fed. R. Evid. 401 & 402.
3. Fed. R. Evid. 801 (a) and (c).
4. Fed. R. Evid. 802.
5. United States v. Russo, 480 F.2d 1228, 1240 (6th Cir. 1973).
6. Fed. R. Evid. 901(a).
7. Fed. R. Evid. 901(b) (10)
8. Comment, Guidelines for the Admissibility of Evidence Generated by Computer for Purposes of Litigation, 15 U. Cal. Davis L. Rev. 951,956 n. 17 (1982) and Henak, Robert R. and Ellen Henak, Using Computer Printouts in the Courtroom, Wisconsin Lawyer, March 1989, p.12
9. "Note: Assuring the Competency of Computer Generated Evidence", 9 Computer Law Journal 103, 105 (1989).
10. Note, p. 105
11. Note, p.109
12. Fed. R. Evid. 803
13. Ibid
14. Henak, p.58

AN INTELLIGENT SOLUTION FOR KINEMATIC GPS MAPPING

J. V. R. Paiva
Vice President, Research & Development
The Lietz Company
9111 Barton
Overland Park, KS 66214
(913) 492-7574, ext. 551

BIOGRAPHICAL SKETCH

Joseph V. R. Paiva is Vice President, Research & Development at The Lietz Company and General Manager of Sokkia Technology, Inc. In this capacity he is involved in the development of software-based products for the surveying and mapping industries. His previous experience includes teaching surveying at the University of Missouri-Columbia and being self-employed as a consulting surveyor and engineer. Dr. Paiva is licensed as a land surveyor and professional engineer. He is a member of ACSM, NSPS, ASCE, NSPE and the Missouri Association of Registered Land Surveyors.

ABSTRACT

GPSMAP is a software product designed to establish an intelligent link between an otherwise-conventional Electronic Field Book (EFB) used with total stations, the 4000ST series GPS receivers and PC-based software used for mapping applications. The EFB software component sets up two-way communication with the GPS receiver, previously set to kinematic mode. Once in the surveyor's backpack, the EFB acts as the interface to the receiver. More importantly, it sends feature code and other information about the point being observed to be stored in the receiver's memory. The EFB software can also monitor the number of satellites being observed and inform the surveyor of receiver battery and memory status. When insufficient satellite signals are being received, the EFB warns the user and advises as to the last point observed to enable restarting. Map editing can be done immediately after the survey is completed, using the data files stored in the EFB because the EFB stores point positions. After post-processing the GPS data, the high-accuracy positions can be read from the post-processing system, and substituted for the point positions recorded in the EFB. This solution provides an elegant option for those wishing to collect position data with descriptive codes, using a single person, for planimetric, contour and GIS mapping applications.

INTRODUCTION

This paper presents a new, practical concept for a solution to the problem of using kinematic GPS to quickly and efficiently produce maps. Given the current size of GPS receivers used in surveying applications requiring centimeter accuracy kinematic surveying, the logistics of carrying receiver and antenna from point to point; collecting sufficient data at each point;

monitoring whether loss of lock occurs or the number of satellites being observed drop below a pre-determined number; and recording information about the point being observed so that its position can be correctly tagged after post-processing, can be a daunting proposition.

Using a common series of data collectors in use for surveying applications involving total stations, and PC software which is sophisticated enough to handle the unique qualities of GPS-derived data, most of the operational problems encountered in ordinary kinematic GPS surveying are eliminated. An added benefit is that once the data has been processed through the PC software, they may then be used to generate contour maps, planimetric maps, ported into AutoCAD, or formatted in a variety of ways to transfer them into a GIS.

GPSMAP is a development by Datacom Software Research Limited in cooperation with Trimble Navigation Limited. The coordination of development, application testing, documentation and production has been coordinated by Sokkia Technology, Inc., a subsidiary of The Lietz Company.

PRINCIPAL COMPONENTS

As shown in figure 1, the chief component of GPSMAP is a set of disks containing a PC program which requires the MAP module of Sokkia Software to be loaded as well. When GPSMAP is purchased, also included are disks for loading software into the SDR33 or SDR20 series Electronic Field Book; and a "Trimble" adaptor. These are described further below.

PC software

Sokkia MAP is an automated survey data processor which also creates a database and plots maps with user defined symbols and line types. In surveying practice with total stations and the SDR33 or SDR20 series Electronic Field Books, the user may enter any system of descriptive codes (feature codes) to describe the points being observed. In addition to using codes such at "tree," "manhole" and "curb," control codes or control codes with parameters may be used. The surveyor may define any set of control codes as well, but some typical ones may be "ST" (for start), SIZE n (for size of tree crown for example, where n would represent the parameter defining either its scaled size or plot symbol size). This type of coding system puts few, if any, limitations on the surveying process. When surveying a road, for instance, the left back of curb line, left gutter, centerline, right gutter, right back of curbline and even fences and sidewalks, may all be surveyed by repeatedly going *across* the street while progressing down, instead of surveying the left fence to the most distant point possible, then starting the left curbline, and so on. Added to these features are data editing; recalculation routines; automatic labels, joining, symbol and line plotting.

GPSMAP, when installed with Sokkia MAP gives MAP additional enhancements. The primary one is that the input files can be in latitude/longitude/height format. It can also receive files of post-processed data produced by Trimvec-Plus.

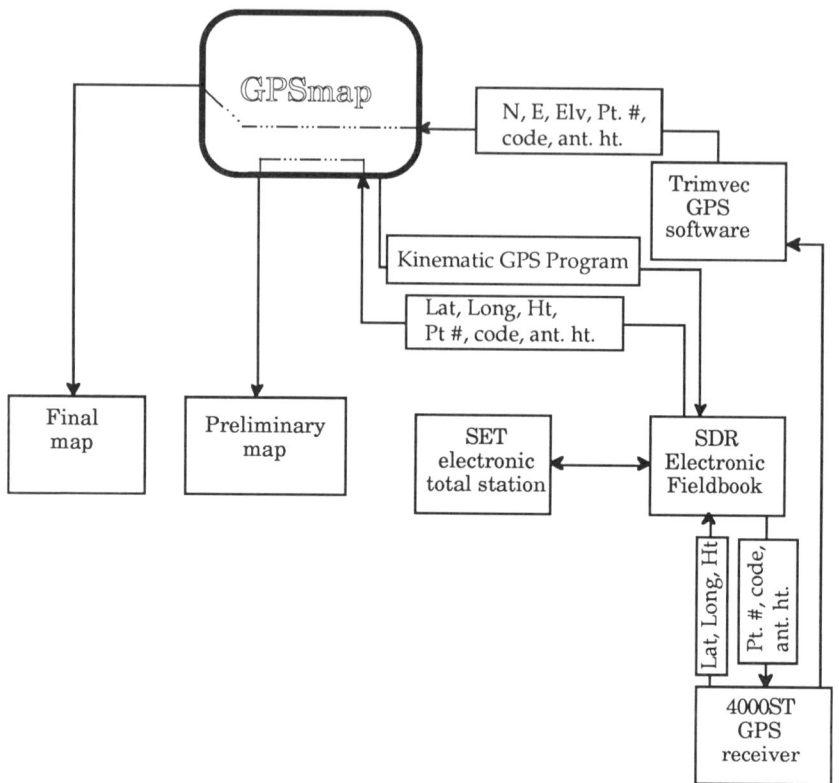

Figure 1: GPSMAP Block Diagram

Electronic Field Book software

GPSMAP contains software for loading a program called "Kinematic GPS" into the SDR33 or SDR20 series of Electronic Field Books. In the case of the SDR33, the new program resides in the EPROM. In the SDR20, because of space limitations, the program occupies RAM space.

When called, this program establishes communication with the Trimble 4000ST series of receivers (if already in kinematic mode), locks the 4000ST's keyboard, and allows the keyboard and display of the SDR to be used to control certain aspects of the GPS receiver and to monitor its status.

GPS receiver firmware

To have two-way communications between the SDR and the 4000ST series receiver, a firmware upgrade is necessary if not already so equiped. The minimum versions required are: boot ROM 2.0, nav. firmware 4.2, sig. v. 4.2.

This firmware allows the SDR to perform many automatic checks of the GPS receiver's status. It also allows, identification data to be sent to the GPS receiver for storage in its memory and point position from the GPS receiver to be stored in the SDR's memory.

OPERATION OF SYSTEM

After installing the "Kinematic GPS" program in the SDR, the kinematic survey is started according to standard GPS procedures. The "Trimble adaptor" is a device which plugs into the data/battery port of the 4000ST series receiver. This is a splitter which enables the battery to be connected to the receiver, as well as the SDR. With the GPS receiver in kinematic mode, it may be placed in a backpack, with cables for antenna and SDR running to a prism pole onto which a kinematic antenna and the SDR as well have been attached.

Start up

When ready to begin the stop and go part of the survey, the "Kinematic GPS" program in the SDR is called. It established communications with the receiver and locks the receiver's keyboard. Then the SDR prompts for receiver and antenna type. The read key is pressed next and the user is prompted for a point number. When entered, the SDR puts the reciever into *static* mode (see Figure 2).

Entering feature codes

Depending on the options selected in the SDR, the prompts for meteorlogical data are next displayed. If the antenna height has changed this may be entered as well. At this stage, the prompt for feature code is also displayed. The surveyor may type in a descriptive note as well as any special control codes to control plotting of the sysmbol or line work after post-processing. The SDR's feature code stack facility may be used. This facility enables a code to be called up out of memory by typing the one or two characters which begin the code. The codes may be pre-entered into the SDR's memory using a keyboard input facility or by downloading a file of codes from a PC.

Monitoring for adequate measurements

Prior to starting the survey, the number of GPS measurements required at each stop should have been set. The SDR tracks the number of measurements since the read key was pressed, and displays this by counting down. Thus when the number of measurements required is zero, the surveyor knows that sufficient measurements have been taken. If the surveyor wishes, additional measurements may be taken, and the countdown continues into the negative numbers.

The point number, antenna height, feature code and meteorological data are stored in the SDR's memory. It is also transmitted to the GPS receiver where it is stored with the GPS measurements just made. When this data is received by the 4000ST, it also sends to the SDR the point position (latitude/longitude/height) which is stored in the SDR's memory (Figure 3).

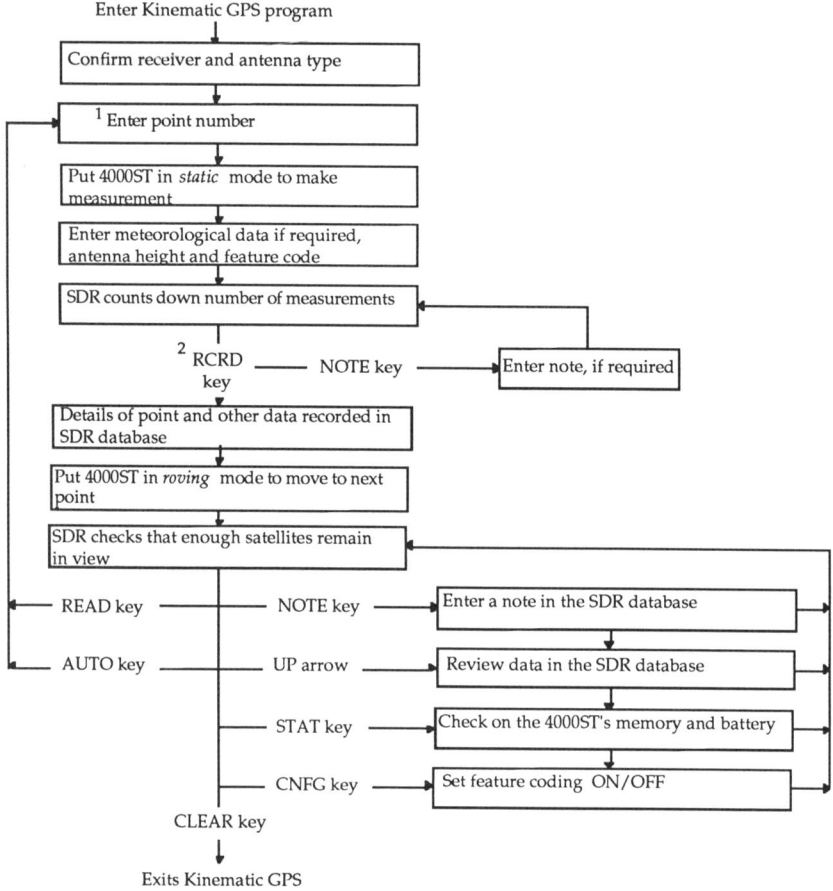

Figure 2: "Kinematic GPS" operation in Electronic Field Book

At anytime, the surveyor may press a key on the SDR to determine the GPS receiver's remaining data storage and battery capacities. The database in the SDR may also be reviewed at any time. Explanatory notes up to 60 characters long may be entered at any time as well.

Restarting after loss of lock

As soon as the data is recorded, the SDR puts the receiver in *roving* mode. It then displays the number of satellites in view as the surveyor walks to the next point to be surveyed. If the number in view drops below a preset number, the SDR beeps and displays information on the last point surveyed in case the surveyor wishes to restart the survey from there. If not, the surveyor may view and scroll through the database (which will show the descriptions entered for the points) to find a reliable point from which to restart. When at the desired point, the SDR prompts for a static observation to do the restart.

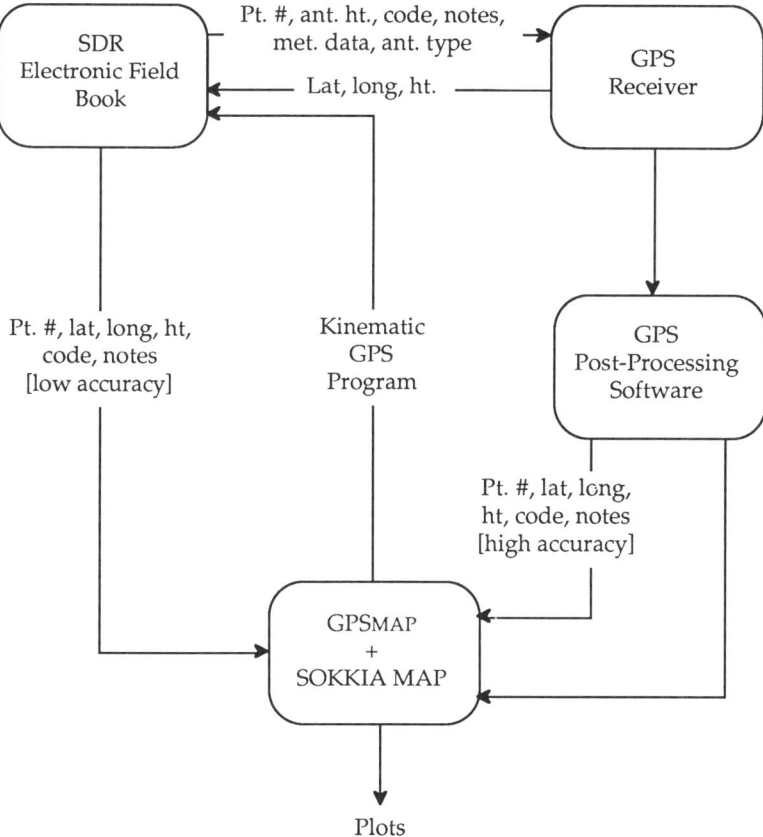

Figure 3: GPSMAP Data Flow

PROCESSING THE DATA

When the survey is complete, the "Kinematic GPS" program is exited which casuses the two-way communciation with the GPS receiver to cease and the 4000ST's keyboard is unlocked.

Data in the SDR may be directly downloaded into the GPSMAP/Sokkia MAP program for processing of the point positions. Data stored in the 4000ST may be post-processed in Trimvec-Plus and the precise positions then tranferred to GPSMAP.

Coarse data

So-called because it contains point positions, "coarse" data may be processed for several reasons. Where it is intended to use precise, post-processed data, the coarse data may still be processed through GPSMAP to generate a plot to determine what editing of codes, lines, etc. are required. This is a valuable feature since post-processing of the GPS measurements may take hours whereas the GPSMAP processing and Sokkia MAP plotting can usually be completed in a few minutes.

Where the survey extends over several days, the "coarse" plots can also be used as a planning tool by management to ensure that the project is being properly surveyed.
Finally, "coarse" plots may be used in certain applications (mining roads, logging roads, various reconnassance surveys) as the end-product.

Because Sokkia MAP contains Lambert Conformal, Transverse Mercator and Mercator projections, the conversion of coarse data to plane coordinate values is easily done.

Precise data

When GPS data has been post-processed by Trimvec-Plus, the output file may be read into GPSMAP in latitude/longitude/height or Northing/Easting/Elevation form. The point numbers and feature codes are appended to the positions since they were stored in the GPS receiver's memory as well.

Once read into GPSMAP, the data may be translated, rotated and scaled. If desired, a height correction plane may be determined using known orthometric heights of GPS points and a height adjustment for every GPS point calculated. Alternatively, a least squares fit to several known points may be calculated.

If coarse data has already been processed, and editing of the map already done, Sokkia MAP's featuring allows easy substitution of the precise position for the coarse ones so that the editing process does not have to be repeated.

At this stage, the data may be used to generate contour maps, profiles and cross sections, etc. It may also be written out in a variety of formats to transfer into many software products including GIS, CAD and engineering design systems.

APPLICATIONS

While the obvious use of GPSMAP is for topographic mapping, other applications can include planimetry, control, cross sectioning, profiles, checking construction, and even hydrographic surveys. Probably the most potential for dramatic impact is when GPSMAP is used as the front end to GIS applications requiring data of surveying quality.

In a recent test of GPSMAP, 74 positions were surveyed at an intersection, first with GPS, then with a total station. A review of the differences in Northing, Easting and Elevation are as follows (Taylor, 1990).

	ΔN	ΔE	ΔElv	ΔDistance
σ	±0.006m	±0.006m	±0.006m	±0.001m
Mean	−0.001m	0.002m	−0.004m	0.011m
Minimum	0.0m	0.0m	0.0m	0.0m
Maximum	0.014m	0.184m	0.143m	0.0198m
Range	0.028m	0.036m	0.025m	--

CONCLUSION

GPSMAP is an innovation which combines a conventional data collection system and PC processing software, both with modifications, with kinematic GPS surveying. It offers the potential for increasing the productivity of GPS surveying; it also makes it possible through the SDR Electronic Field Book and Sokkia MAP software, to integrate data from conventional (total station) surveys with GPS-measured data.

Training and use of the system should be a relatively straightforward matter for many surveyors. The SDR Electronic Field Book series is the most popular data collector used in the U.S. surveying market (P.O.B., 1990), thus the possibility that the surveyor is already familiar with it is high. Other potential advantages include a lower investment if the data collector is already owned, as well as the possibility that less training will be required.

REFERENCES

P.O.B. "P.O.B. 1989 Equipment Survey," April/May 1990, V. 15, No. 4, pp. 78-80.

Taylor, Arthur. Private Communication, December 28, 1990.

THE OWENS VALLEY GPS CONTROL SURVEY

W.W. Parks
J.K. Crossfield
B.S. Littell

ABSTRACT

The project to collect and analyze and adjust GPS data for the Los Angeles Department of Water and Power in Owens Valley California is described. Data collection is described. Baseline processing and network analysis and adjustment are described. Project planning is described in a separate paper. Single frequency and dual frequency data were processed. Single frequency processing is illustrated by an example. Results of processing 528 baselines by various methods is described. Network analysis by minimally constrained adjustment is described and suggests that the network exceeds criteria for first order geometric classification. Constrained adjustment to estimate coordinates of network stations is described.

PROGRESS REPORT
ON THE
DEVELOPMENT
OF AN
INTEGRATED PLSS
CADASTRAL MEASUREMENT MANAGEMENT
AND
RETRACEMENT SURVEY SOFTWARE SYSTEM

Corwyn J. Rodine
Bureau of Land Management
Eastern States Office

Jerry L. Wahl
Bureau of Land Management
California State Office

Blair Parker
Barry M. Blanchard
Raymond J. Hintz
University of Maine

ABSTRACT

Cooperative work between Cadastral Survey of the Bureau of Land Management and the University of Maine has resulted in development of an integrated Measurement Management software system specific to retracement cadastral surveys in the U.S. Public Land Survey System. The software is now in a beta release phase. This paper will provide some background for the unique application requirements and a progress report on the system development. The paper will provide descriptions of many of the capabilities of the system, and a brief discussion of the many challenges that have faced this development.

INTRODUCTION

Cadastral Measurement Management (CMM) is an ongoing software development effort between the Bureau of Land Management and the University of Maine which began approximately 2 years ago. The beta version of CMM has undergone a series of tests within BLM. The result is a system of integrated programs which are specifically designed for dependent resurveys within the U.S. Public Land Survey System (PLSS).

Base design considerations of CMM are:

(1) Geodetically correct PLSS computations are provided at all program stages.

(2) All computations are in strict conformance to the Manual of Surveying Instructions, 1973.

(3) All survey measurements will be automatically subjected to least squares analysis in a defined geodetic framework. The software will not inhibit the way in which one surveys or enters data.

(4) The user will have all "manual" computational tools at his/her disposal, or highly automated batch processing of a series of computations.

The development of CMM was deemed necessary because no available software package on a personal computer met the four base design criteria.

Development and Testing Strategy

While the software development was performed primarily at the University of Maine, the Bureau of Land Management had a great deal of interaction in its development through the Cadastral Technical Advisory Group (CTAG). This group was selected based on their experience with automation in cadastral surveying, and consists of Corwyn Rodine (Eastern States Office), Jerry Wahl (California State Office), Tom Noble (Denver Service Center), Tom Wohlend (Alaska State Office), and Bernard Hostrop of the Washington Office. CTAG meets at the University of Maine at least twice a year for one week time periods, and interacts with University personnel on a regular basis as updates to software are provided and tested.

To perform primary testing of the alpha version of the software, a primary test site was needed. Each BLM state office interested in performing the beta testing was required to submit a proposal detailing why they could provide an effective testing mechanism. Offices could also request to serve as secondary test sites. The role of the secondary sites were to provide verification of findings of the primary site.

Though the selection of the primary test site was difficult due to several fine proposals, the Montana State Office was awarded this role. The software was demonstrated to BLM personnel at the March 1990 ACSM-ASPRS meeting in Denver, and during the following week CTAG and University personnel provided initial training to surveyors at the primary test site. Several other state offices also had representatives at this week of training. It was their role to return the software to their office and serve as a secondary test site.

Surveyors at the Montana State Office spent the next two months processing existing data sets through CMM. Each week a status report was provided to CTAG and all BLM cadastral surveyors involved in alpha testing. This essential user feedback was accomplished through the use of the Bureau's electronic mail system. This information was then discussed with University personnel with specific regard to program modifications and further documentation needs.

A follow-up visit to Montana by CTAG and university personnel was made two months after the initial visit. This one week period allowed more thorough "question and answer" sessions to be held since Montana personnel were now users instead of beginners in processing data with the software. This meeting also enabled better forming of the final test report by the Montana State Office.

During the summer of 1990 a series of program modifications resulted from this initial testing, and a more extensive user's manual was prepared. This resulted in the beta release of the software within BLM Cadastral in September 1990.

A CTAG meeting in December 1990 resulted in final required updates before CMM ver. 1.0 is released at the 1991 ACSM-ASPRS meeting in Baltimore.

The extensive interaction between University and CTAG, along with the excellent testing provided by the primary and secondary test sites, has resulted in a very beneficial relationship (Hintz and Rodine, 1990; Blanchard, 1990). It has also resulted in a software system which has been extensively tested and modified for user needs.

Use of CMM in a Dependent Cadastral Resurvey

Getting Started. While every dependent resurvey is unique and therefore cannot be generically categorized, an attempt is made here to describe how a survey would proceed using CMM.

While geodetic positions for PLSS corners and traverse stations are computed using CMM, it is important to document that these are a necessary by-product of the process (Hintz, et al, 1988; Hintz and Onsrud, 1990). Survey measurements are preserved in the system, and as more measurements are added, positions of points are updated. The relative geodetic relation of points is of critical importance as opposed to absolute position.

The first problem confronted by the surveyor is location of geodetic control near the dependent resurvey project area. Since some surveys could be in areas essentially devoid of control, the use of a "scaled" position from a quad sheet will allow CMM to apply correct geodetic analysis to the data in the aforementioned relative relationship. If a survey will eventually tie to geodetic control, it is possible to work from a scaled position until the survey network ties to the control point. At that time the scaled position is easily eliminated as a control point.

While not critical in use of CMM, a surveyor will usually find it beneficial to create a digital copy of the official record survey information. To enhance automated computational procedures, CMM can make use of a standard corner identification number which conforms to BLM's Geographic

Coordinate Data Base GCDB. A program called INREC (INput of RECord information) enables efficient input of the record information, and affords the user with a number of testing procedures for correctness of the entered data. The correctness of the digital record information is of utmost importance in the dependent resurvey, and the thorough checking of its validity cannot be overemphasized.

CMM Data Analysis

Data Entry. In an ideal scenario, collected field data should be input to CMM daily. This philosophy can easily be overridden by busy field days and other duties, but their are several reasons why entry on a daily basis is useful.

The testing by the Montana State Office showed that data entry through keyboard input was very monotonous and error prone. This was especially true in the testing since complete jobs were being entered at one time. Entry of a small amount of data (such as a day's work), followed by verification of the data through the analysis routines in CMM, enables the user to look for problems in a finite amount of data if the previous data has been verified.

The best solution to the data entry problem will be direct input from field data collectors. This has already been accomplished for existing data collectors being used within BLM, and will be further resolved with the development of the BLM Cadastral Electronic Field Book (Wahl, et al., 1991).

Keyboard entry of data is accomplished through a dedicated entry/editor system within a CMM program called GENER. GENER's other critical role is coordinate generation in a conventional coordinate geometry sense. For dependent resurveys the coordinate generation is almost entirely from traverse type computations. GENER requires no specific ordering of data, and thus traverse routes with closure reports are generated for any data order. The closure reports are extremely useful in blunder detection prior to least squares analysis. GENER also allows generation of only new station coordinates, thus preserving existing adjusted values if desired. GENER provides a multitude of error messages for items such as inability to generate coordinates for all stations.

If the data appears blunder-free the most important role of GENER is that approximate coordinates have now been automatically generated for any traverse network, and these approximations fuel the ensuing parametric least squares analysis.

Least Squares Analysis: The successful use of least squares analysis of dependent resurvey data has been discussed in *Hintz and Rodine* (1989). One of the critical items in acceptance of CMM has been the demonstration that data sets processed thus far reflect statistically insignificant amounts of adjustment to measurements. The use of least squares has thus become a process of verification (analysis) that the geometric constraints of redundant data do not require unrealistic adjustment of any field measurements. Since resulting coordinates are geodetic in nature the software has the ability to handle meridian convergence and projection scale factors automatically, and elevation information can be provided by the user at a level of complexity ranging from a single project elevation to individual elevations for each traverse station.

The user has the ability to assign project default error estimates, and override any of these with individual measurement values or a similar estimate for a series of measurements. The latter being critical in the execution of a Dependent Resurvey, because in "following the footsteps of the original surveyor", one often finds oneself in stretches of nasty terrain such as swamps and bogs. One of the most important items noted in testing is verification of the surveyor's ability to estimate his own measurement error. Error ellipse information of final positions are also available to the user.

Users also have the ability to use the least squares analysis in blunder detection through residual examination and standard robustness techniques. A user must be aware that these technique's successes rely heavily on the geometry and redundancy of the survey network.

The final important item in the least squares analysis was the ability to adjust very large networks on a personal computer. The beta version release of 3000 stations has already been exceeded by a job in Arkansas which consisted of an entire township with significant section subdivision work, and thus a version capable of larger adjustments will be available for these special cases in release 1.0 of CMM.

Traverse Reports: Since the least squares analysis does not provide conventional traverse closures a program called CHECKER performs this function automatically. Conventional latitude and departure closures, linear precisions, and angular error of closure are generated both in an automated fashion and by user definition. CHECKER has also been used extensively for blunder detection purposes since it reflects the amount of adjustment which has been applied to a series of measurements.

Computerized Viewing of Survey Network: Users are able to obtain graphical representations of their survey network using two separate methods. A quick view of the survey network is available if the PC is equipped with a graphics adaptor using a program called VIEW. This program has recently been modified to include pan and zooming operations, allowing users to view any portion of their survey network very rapidly. CMM also has the ability to create a standard DXF file for viewing with CAD software. The later option is also being used for input of information to CAD for final plat preparation.

CMM Cadastral Computations

The aforementioned analysis routines can be utilized for any type of surveying, and thus are not restricted to dependent resurveys or retracements within the PLSS system. The following cadastral computations are, for the most part, internal to PLSS computations.

Manual Proportioning using the record file: The traverse network eventually provides suitable information to enable single and/or double proportionate measurement for search locations or for recomputation of corners deemed lost. The user has the ability to identify a corner, and computer program PROPORT automatically searches for controlling corner information and necessary record information. If this information is available the computation is performed according to the Manual of Surveying Instructions, 1973 and a report is created which details this information. The report can be stored as a file for later referral as to the order and basis of computations which the surveyor performed.

Automatic Proportioning using the record file: If GCDB point identifiers are used, CMM has the ability to analyze controlling positions and the record file, and automatically proportion the entire data set. This ability is extremely useful in determining search positions for corners, and at the end of a job can be used for final determination of all positions of corners which have been identified as lost. A significant report is generated so the user can thoroughly check this automated process. The program is also being used as an aid in evaluating the reliability of local monumentation as one can compare where they are located relative to a proportionate position from found original corners.

Geodetic COGO (CSTUF): One of the most important tools for the cadastral surveyor is the ability to perform coordinate geometry computations in a geodetic framework. With this program the user can add, delete, and list coordinates in either geodetic or cartesian coordinate (plane) mode. The conventional coordinate geometry operations of bearing-bearing intersection, distance-distance intersection, distance-bearing intersection, coordinate inversing and traverse computations can be performed in either geodetic or plane mode, and any geodetic bearings can be identified as forward or mean in nature. Geodetic and plane midpoint round out the computational options along with area computations. Basic horizontal curve computations are being added in ver. 1.0.

Viewing of PLSS information: The user has the ability to graphically illustrate what corners have been positioned, or where temporary corner positions have been located. This is performed through use of a computer-screen version of a township diagram with graphical identifiers for all information.

Corner moves and true line offsets: These functions are analogous to layout in a dependent resurvey sense. A corner move returns a set of angle and distance combinations from existing stations which identify a geodetic position. This function could be used for setting both temporary and final corner positions. Once true line has been identified through retracement or restoration, true line offsets in the form of angle or bearing and distance can automatically be generated from all traverse stations within a user-defined distance from the true line.

Miscellaneous utilities: This program enables proportioning without use of the digital record file. This is also the program where "adjustments" within the Manual of Surveying Instructions such as broken boundary, irregular boundary, etc. are performed. This program also has the ability of performing one, two, or three point control computations in accordance with the Manual of Surveying Instructions, 1973.

Relation of Data Analysis to Cadastral Computations

It is important to recognize that the two components of CMM are used in unison on a continual basis. Survey measurements are entered, analyzed, and verified. This additional information is used to update search positions for corner locations. As monuments are found and identified these corner locations are used as controlling information in ensuing cadastral computations.

The survey measurements are thus used in providing better information in the dependent resurvey process. Each job will be unique in how information is collected and utilized.

SUMMARY

A cadastral measurement management software system has been developed through co-operation between the Bureau of Land Management and the University of Maine. The software system is flexible in that all of the computational complexities of the U.S. Public Land Survey System can be resolved in a geodetically correct fashion. The software has been placed through extensive test procedures and revisions in the realization that only this approach can result in a system which will be useable by the general surveying community.

Acknowledgements

The authors wish to especially thank Steve Douglas, Mark Dixon, Dan Mates, and Steve Toth of BLM for support in software testing, Scot MacDonald of the University of Maine for contributions to the software development, and the support of other CTAG members Bernard Hostrop, Tom Noble, and Tom Wohlend.

References

Blanchard, B.M. (1990), *Utility Program Development: A Digitally Integrated Measurement Management System for the U.S. Public Land Survey System*, M.S. thesis, University of Maine, Orono, ME, 86 p.

Hintz, R.J., Blackham, W.J., Dana, B.M., and J.M. Kang (1988), *Least Squares Analysis in Temporal Coordinate and Measurement Management*, Surveying and Mapping, Vol. 48, No. 3, pp. 173-183.

Hintz, R.J. and C.J. Rodine (1990), *Automation and Precision in a Cadastral Surveying Environment*, Proceedings of the ACSM-ASPRS Annual Convention, pp. 124-133.

Hintz, R.J. and H.J. Onsrud (1990), *Upgrading Real Property Information in a GIS*, URISA, Vol. 2, No. 1, pp. 2-10.

U.S. Dept. of Interior, Bureau of Land Management (1973), *Manual of Instructions for the Survey of the Public Lands of the United States 1973*, U.S. Government Printing Office, 333p.

Wahl, J.L. (1990), *Cadastral Measurement Management User's Manual (Beta Release)*, unpublished, 120p.

Wahl, J.L., Rodine, C.J., and R.J. Hintz (1991), *Development of Electronic Field Book for Cadastral Retracement Surveys*, Proceedings of the ACSM-ASPRS Annual Convention, accepted for publication.

A GRASS ROOTS STRATEGY FOR ESTABLISHING THE SURVEYING PROFESSIONAL'S PROPER ROLE VIS-A-VIS LIS
Gene V. Roe, Ph.D., LLS

ABSTRACT

Nearly fifteen years ago, as a newly licensed land surveyor, the author became aware of the concept of a multipurpose cadastre, and along with it the promised benefit of improved land record keeping systems. Today, the U.S. surveying profession appears no closer to setting the national agenda for such developments, let alone delivering any tangible results.

Coincidentally, the U.S. surveying community continues to seek a professional stature on a par with their counterparts in other advanced countries of the world. This paper will issue the surveying professional a challenge. A challenge to develop a mandate at the local level, whereby the professional surveying community becomes the custodian of the parcel layer of each LIS, and ultimately demonstrates to all vested interests that they are being treated more equitably as a result. If this can be accomplished the rewards will follow.

INTRODUCTION

"The restructuring of America from an industrial
to an information society will easily be as pro-
found as the shift from an agricultural to an
industrial society."
John Naisbett
(Megatrends,1984)

During the 1980's the surveying profession was the beneficiary of truly revolutionary advances in the tools and technology made available to them. Total stations, data collectors, and desktop computing systems all became available at an affordable price in just a brief 10 year period. Soon the most important technological breakthrough in the history of surveying - GPS - the Global Positioning System, will also become a cost effective tool for the local surveyor.

On another front, Geographic Information System (GIS) technology was also exploding onto the surveying scene during the 1980's. Digital land information databases are now currently being assembled at all levels of government, and by the private sector as well. This powerful technology is the final element in the total shift from the hands-on, analogue world of surveying that was commonplace in 1980, to the black box, digital world of land information systems (LIS) of the 1990's. It's no wonder our profession is finding it challenging to cope with this degree of change.

On the other hand, it could be argued that the surveying professional now has all of the tools required to effectively compete in the rapidly changing, information- driven world that we find ourselves living in. The concern that this author has is whether we will seize this truly unique opportunity and move the surveying profession into the 21st Century as one of the leaders of the Age of Information, or whether the U.S. surveying community is destined to be relegated to its second class status for the foreseeable future.

This paper will attempt to describe a simple, grassroots strategy which will insure that the local surveyor is needed to provide key parts of the land information puzzle. That in fact our role be mandated in such a way that we are able to leverage the need for our services into a higher regard for the skills and talents of our profession. By becoming the custodian of the geodetic control and cadastral information layers of the local LIS, we will be in a position to demonstrate to the public that, in the long run, they will be treated more fairly and equitably. These three concepts, as identified by Chrisman (Chrisman, 1987), are essential to the widespread acceptance and assimilation of LIS technology into the management of land related activities.

INFORMATION SYSTEMS

Being in the land information business should really be nothing new to the surveying community - that's the business we have been in since the beginning. From the ancient Egyptian "rope stretchers", to the present day land information specialist, the surveyor has been measuring, recording, and portraying his/her results as they carry out the mission which the public has entrusted them with. Our business has always been information based, unlike many other sectors of the economy.

What is different, of course, is the shift in technique brought on by the introduction of the microprocessor to virtually all aspects of the surveyor's tasks. From the data collector, to the total station, to the desk top computer, and finally the GPS receiver todays land surveyor must operate in a digital world.

Along with the introduction of our new high tech tools, has also come the systems approach to the organization of our thinking. As seen in Fig. 1, McGlaughlin and Nichols prepared a hierarchy, or taxonomy of information systems as they apply to the world of geographic location (McGlaughlin and Nichols, 1987). Working down through this tree structure, we want to focus on the block labelled, "parcel- based information systems". This is the type of LIS that this author is identifying as the information system which the local surveying community should be seeking to establish throughout the U.S. This is the scale of LIS that this author believes was recommended by the two National Science Foundation reports in the early 1980's on what was called, at that time, the "multipurpose cadastre".

For some reason all of the hard work that went into these reports has seemingly been forgotten, or is being ignored by the surveying organizations, as other professional groups, such as URISA, are making their case for being the leaders of the GIS/LIS movement. If you have not reviewed these reports, this would be an important activity, for they clearly lay a foundation for the critical role which the local surveyor is best suited to play in the development of parcel-based LIS.

INNOVATION AND CHANGE

Let's assume, for the purpose of this discussion, that the surveying profession could agree to support the concept of a parcel-based, LIS implementation strategy. How could we best go about achieving the goal of mandating this strategy throughout the U.S.? In this author's opinion, the key lies in a two part strategy.

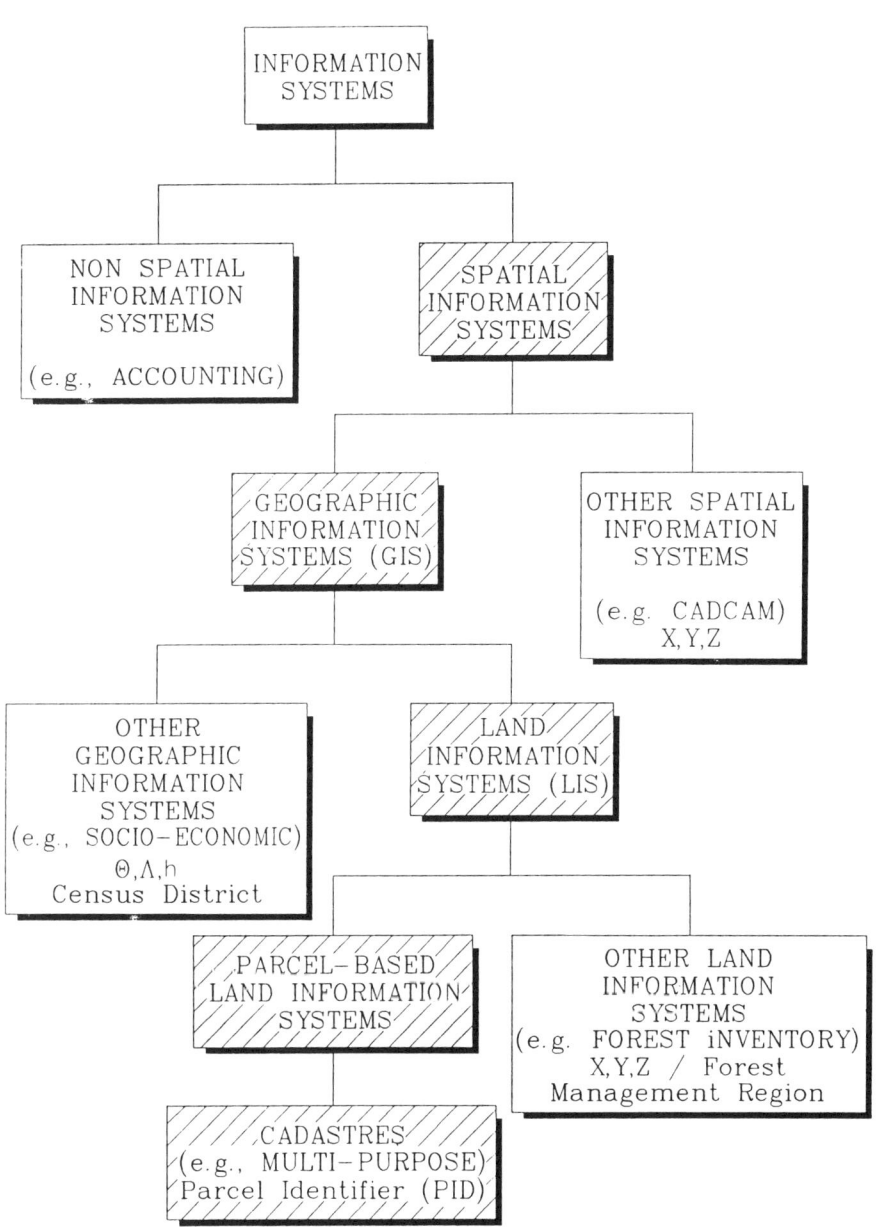

Fig. 1 Spatial Information Systems Taxonomy
(McGlaughlin and Nichols)

The first part would involve setting the national organizations' and state society agendas to focus on establishing goals, effectively communicating these, and then providing the necessary leadership to see the LIS implementation through to successful completion.

This would require a total effort on the part of one or more full time employees that are dedicated 100 percent to this mission. A highly visible, proactive campaign to first educate, and then develop legislative mandates to support the LIS implementation strategy would have to be undertaken.

The second part of the strategy would involve the local surveyor. He or she would be charged with developing the grass roots support for the implementation. This would involve working with the local governing bodies, the legal profession, the banking community, and others to first educate and then, as above, develop the necessary mandates to require changes from the status quo.

Sounds simple enough, but as we all know change and innovation are not that easily brought about. Particularly since the changes being proposed, from analogue to digital, are so radical in nature.

Radical change (revolution), as opposed to innovation (evolution), is not viewed favorably by the establishment. Whether it's the free market system in Eastern Europe or GIS/LIS technology to the surveying profession, the leadership, in general, does not want to see change occur. This is human nature, and it must be recognized and understood so that it can be properly dealt with.

de Neufville and Croissant provide an enlightening discussion of the topic of change and leadership as it applies to GIS technology (de Neufville and Croissant 1990). They point out the tremendous benefits that this new technology has delivered to the cartographic profession, but they are also very concerned about certain trends which they see developing. These trends, resulting from the commercialization of the technology by forces outside the profession, could lead to the death of cartography as we know it today. In this author's opinion, the same arguments could be made about the land surveying profession.

Unless each and every one of us in the land surveying community accepts the responsibility for doing his/her part to bring about the inevitable changes from within our profession, then we are destined to remain in the second class, support role which we now find ourselves in. On the other hand if the surveying profession, led by the local surveyor, were to: 1) accept the challenge of creating a vision of where we want to be in the year 2000, vis-a-vis land information systems, and then 2) motivate people to that end, we could position ourselves in a primary leadership role, not unlike our counterparts in other advanced countries.

THE STRATEGY

Unfortunately, many of us have heard this basic message before. The real problem lies in making contact with those who will not be reading this paper, or attending this conference, because they are not even members of their state society, let alone ACSM. Somehow we have to get their attention, but in many ways that is the subject of a different discussion.

Let's assume for the moment therefore, that we have the critical mass necessary to get the program started, and that if we can generate some early wins that we will be able to develop the true grassroots support that will really make it a success. As introduced earlier the plan for establishing the surveyor's role in relation to land information systems is based on a two level strategy.

The first part involves the national and state organizations. We must rely on them to provide the overall strategic planning and leadership. It should be their responsibility to first create a vision of where we want the surveying profession to be in say the year 2,000. From there we need a mission statement accompanied by the requisite goals, objectives and action plans. This information must then be communicated to the masses. Since every practicing land surveyor must be registered in the state in which they practice, the mailing list is easily obtained for an on-going, direct mail campaign.

Communication and education are the key concepts here, but in the final analysis the most important idea is to convince people to change their focus from where they are, to where they want to be. This critical visualization process of the end result, rather than the process of getting there is essential to creating the environment where the kind of radical changes discussed earlier can be readily accepted.

The second part of the strategy is focused on the local land surveyor. With the vision clearly established on where the profession wants to be in the year 2000, it will be up to the local land surveyor to develop the grass roots demand and mandate for insuring that they become the custodian of the geodetic control and parcel layers of the LIS.

In "Procedures and Standards For A Multipurpose Cadastre" a number of examples are given where communities have begun to develop these mandates, and that was in the early 1980's! (Panel on a Multipurpose Cadastre, 1983) Although the best solution is the one that will work in your community, perhaps the one that will be the most politically acceptable at first is the institution of a requirement for all land development projects that the parcel boundaries be accurately tied to the state plane coordinate system. Over a ten to twenty year time period in those areas where development is occurring, or about to occur, these newly developed tracts will be precisely "locked in" to the local LIS. Once the benefits of this are clearly demonstrated through the more equitable treatment of the taxpayers of the community, a more aggressive program can be undertaken to retroactively capture the remaining areas. This is obviously a long term program, but one that must be started NOW!

This small, but extremely important requirement of tying property boundaries into the state plane coordinate system is the mandate needed to place the local surveyor in a position of being the custodian of the control and cadastral layers of the LIS. Without GPS this requirement would have been very difficult to sell, even to the surveying community. With a full satellite constellation the actual work required, and therefore the cost, will be easily offset by the resulting benefits.

The obvious concern that is certain to arise from this proposal is whether the local surveyor can be relied on to properly complete the GPS tie-in. Although certainly a valid concern in the early stages of the implementation, as time proceeds the knowledge of GPS will become much more commonplace. In any case, overall quality control is a function that can be simply monitored by a central regional agency, such as a county government, or through a contract service provided by a private firm. The latter could perhaps be organized initially by the state land surveying society. Their assistance and support is obviously critical to the success of this strategy as they could be the source of both technical training in GPS and LIS technology, as well the catalyst for political change.

CONCLUSION

There is no question that many of the details of the implementation strategy have not been presented here. They represent the process. The intent of this paper is to develop a vision of where we want to be in 10 to 20 years as a profession, and place the focus of our attention there - the details will take care of themselves. Trust me.

As the last decade of the twentieth century begins, the time is now for the land surveying profession to be heard. We must not allow other groups to dictate the accuracy requirements of the control and cadastral layers of the LIS. Tax map accuracy is not sufficient as a basis for a local LIS. Sure the costs are very high to "do it right the first time", but that's the way it usually is. If we do not speak up now, we will have no one to blame but ourselves for missing this truly unique opportunity to change the way in which the land surveyor is viewed by the other professional communities in the United States.

REFERENCES

Chrisman, N.R., October 1987, Design of Geographic Information Systems Based on Social and Cultural Goals, Photogrammetric Engineering and Remote Sensing, Volume 53, pp 1367-1370.

de Neufville, R. and Croissant, J., Fall 1990, A Policy for Technology Leadership, Journal of the Urban and Regional Information System Association, Volume 2, Number 2, pp 7-15.

McGlaughlin, J.D., and Nichols, S.E. March 1987, Parcel- Based Land Information Systems, Surveying and Mapping, pp 11-27.

Naisbitt, J., 1984, Megatrends, Warner Books, p. 9.

Panel on a Multipurpose Cadastre, 1983, Procedures and Standards For A Multipurpose Cadastre, National Academy Press.

INTEGRATING GPS INTO THE
BUREAU OF LAND MANAGEMENT's
CADASTRAL SURVEY PROGRAM

Robert Scruggs
Bureau of Land Management
Division of Cadastral Survey
1849 C Street, NW, Premier 201
Washington, D.C. 20240

ABSTRACT

The Bureau of Land Management (BLM) Cadastral Survey Program began using the Global Positioning System (GPS) in 1988. Before that time, Doppler satellite positioning and terrestrial geodetic surveying techniques were employed by BLM cadastral surveyors to establish geographic coordinates on selected corners of the Public Land Survey System (PLSS). Presently (December 1990), GPS is being used by BLM land surveyors to establish control on selected corners of the PLSS, to assist in the location of PLSS monuments, and in the development of the Geographic Coordinate Data Base (GCDB). In addition, field testing is currently being conducted to examine the feasibility of using GPS to perform cadastral surveys and resurveys.

INTRODUCTION

The BLM Cadastral Survey Program purchased seventeen Motorola Eagle GPS receivers in 1988. Eleven receivers are maintained and managed in a centralized location by the Cadastral Equipment Cache for use by BLM's twelve State Cadastral Offices. The other six receivers were purchased by the California State Office, the Oregon State Office, and the Wyoming State Office. In addition to the seventeen Motorola Eagle receivers used for conducting control surveys, the Cadastral Equipment Cache also manages four Trimble Pathfinder receivers and two Magellan receivers that are used for navigational applications.

In 1973, the BLM Cadastral Survey Program began using Doppler satellite position techniques to establish control on PLSS corners. Approximately 1650 Doppler positions were established between 1973 and 1988. The use of Doppler ended in 1988 when GPS receivers were first acquired by BLM. Since 1988, GPS has been used to establish control on over 2500 PLSS monuments.

Typically, control established on PLSS monuments consists of latitude, longitude and height. Height, although routinely measured, is generally not established at the same level of accuracy as latitude or longitude. This can be attributed to two factors; (1) existing horizontal control stations lack precise heights, and (2) height is generally not required for most cadastral survey applications.

Cadastral surveyors use GPS as the preferred method of determining latitude and longitude of selected PLSS monuments during the course of a cadastral survey or resurvey. The latitude and longitude of these monuments are recorded in the official cadastral survey field notes. Also contained in the field notes are the names, descriptions, and coordinates of the controlling horizontal control stations.

Advances in Geographic Information System (GIS) and Land Information System (LIS) technologies have focused attention on the use of GPS. GPS is seen by many as an economical way of collecting geodetic quality data for GIS/LIS applications. For the most part, GPS is used in GIS/LIS applications to strengthen base map control. As the sole Federal agency responsible for the maintenance of the PLSS on all Federal lands, BLM is increasingly being asked to furnish GPS data on PLSS corners to other government agencies, the private sector, and individuals.

GEOGRAPHIC COORDINATES AND CADASTRAL SURVEY

<u>Brief History</u>

The need for geographic coordinates has not changed substantially in the last 200 years. Beginning with the Land Ordinance of 1785, geographic coordinates have been established on selected corners of the PLSS. Coordinates were first determined on selected points of a survey as a method of checking the alignment of principal meridians and standard parallels. The coordinates, determined by astronomical methods, were used to check lines that were originally established with an ordinary compass (magnetic).

With the invention of the solar compass in 1835 by William A. Burt, U.S. Deputy Surveyor, the needed for geographic coordinates took on additional importance. The solar compass allowed surveyors to determine direction independently of the ordinary compass which had been a constant source of error. But unlike the ordinary compass, the Burt solar compass required that the latitude be known in order to determine direction. Since there were very few maps available in which the latitude could be scaled (and even fewer horizontal control stations in which the latitude could be directly measured), the surveyor first had to determine the latitude using astronomical methods prior to commencing the survey.

The Manual of Surveying Instruction (Manual), 1894, was the first Manual to recognize the placement of geographic coordinates on survey plats. The Manuals of 1855, 1881, and 1890 did not mention the use of geographic coordinates. By depicting geographic coordinates on the survey plats, local surveyors (who were generally contracted to subdivide the township or sections of the township) did not have to redetermine the latitude for use with their solar compasses or transits with solar attachments prior to commencing their surveys. This saved both time and money.

The 1902 Manual did not contain additional information regarding the use of geographic coordinates nor did the 1930 Manual. However, the 1930 Manual does contain a substantial amount of information on how to determine time, azimuth, and latitude.

The 1947 Manual is the first manual to specifically address reasons for establishing geographic coordinates on PLSS corners, apart from the now topical reasons of determining time, azimuth, and latitude. Section 133 reads: "In practical application ... the usual purpose is to ascertain a latitude and longitude for the monument of the public land survey that may be sufficiently good for mapping, reference, and other ordinary use excluding significance as a geodetic control station."

The 1973 Manual was issued prior to the BLM becoming involved in GIS/LIS applications. This Manual does not address the use of geographic coordinates in GIS/LIS applications, but merely restates traditional geographic coordinate usage.

In the early 1980's, the Division of Cadastral Survey issued several Instruction Memorandums regarding the establishment of geographic coordinates on each cadastral survey containing an area in excess of 640 acres. The release of these Instruction Memorandums can be directly linked to the emergence of the LIS concept which was receiving national attention.

Cadastral Control Surveys: Definition

A Cadastral Control Survey (CCS) is defined as a control survey whose primary function is to establish geographic coordinates on selected corners of the PLSS. These surveys are made to transfer the mathematical definition of the earth's surface, represented by the National Geodetic Reference System (NGRS), to legal corners of the PLSS in order to facilitate mapping and other Land Information System functions. A CCS is not intended to extend the NGRS. (Fiedler, 1989, Standards and Specifications for Cadastral Control Surveys)

Purpose of Standards

The primary purpose of a survey standard is to provide a uniform criterion for evaluating results. The specification is the methodology required to meet the standard. Therefore, if all the specifications and other intermediate guidelines are followed, the survey should meet the standard. (Fiedler, 1989) A secondary purpose of a survey standard is to convey to users the quality of the data obtained from a properly executed survey. Presently, the quality of GPS positions can vary between a few centimeters and several hundred meters. The actual accuracy obtained is dependent on the hardware (i.e. type of equipment), software, and methodology. Regardless of the brand of hardware, software, or methodology used, the quality of the results should be expressed in generally accepted terms, i.e., distance standard error, absolute or relative error ellipses, etc. The convention used to convey GPS data quality must be consistent BLM-wide.

Existing Standards for Cadastral Control Survey

In April 1986, the Branch of Cadastral Survey Development prepared the Doppler Handbook for [the] Magnavox MX 1502. This was followed by Standards and Specifications for Cadastral Control Surveys, Conventional Instrumentation in March of 1989. Although the use of Doppler has diminished in the past few years, the Doppler standards still contain useful information that is relevant to GPS. Both the Doppler and conventional standards remain in effect and are to be used by BLM survey employees when conducting Cadastral Control Surveys utilizing these two techniques.

GPS Standards for Cadastral Control Surveys

In August, 1990, the BLM issued Instruction Memorandum 90-567, Policy on the Use of Global Positioning System. These standards are to be used by all BLM survey personnel when conducting Cadastral Control Surveys utilizing GPS.

The Standards for GPS Cadastral Control Surveys provide two GPS classifications, class A and class B, for static, relative positioning techniques. The classes are distinguished by their accuracy and redundancy requirements. The decision on which class to use is dependent upon local conditions and is to be determined by each BLM state office.

GPS Standards and New Applications

As the Cadastral Survey Program becomes more involved in other GPS applications, the standards can be easily broadened to include other GPS positioning techniques. This can be accomplished very easily by tightening or relaxing the classification constant. Further testing will need to be completed before realistic classification constants can be determined and integrated into existing policy. Currently, the BLM is examining GPS standards for cadastral surveys and resurveys.

Cadastral Surveys Using GPS

The BLM, California State Office, Branch of Cadastral Survey, is currently testing the use of GPS in conjunction with cadastral surveys and resurveys. The objective of their test is to compare the accuracy and efficiency of GPS with conventional cadastral survey methods. From this test, GPS standards for cadastral surveys and resurveys can be tested and verified. The final report on this test should be completed later this year (1991).

Distribution of GPS Data

Preliminary GPS data is a term used to describe the status of all GPS data collected in the course of an official cadastral survey. Once a cadastral survey is approved, that portion of the GPS data included in the field notes or depicted on the plat becomes part of the official public record. All GPS data not included as part of the official public record remains preliminary or non-public. The term "non-public records" is defined in IM 90-457, dated May 1, 1990 as:

> All other BLM records such as internal administrative records or those records that must be reviewed prior to release because: 1) a FOIA [Freedom of Information Act] exemption may apply to all or portions of the record; or 2) the Privacy Act is applicable.

Cadastral Survey Field Tablets, Unapproved Cadastral Survey Records (e.g., draft plats and field notes), and Mineral Surveyor Appointment Files are examples of non-public records. Preliminary and non-public GPS records obtained in conjunction with an official cadastral survey may be released to the private sector only under a FOIA request unless an exemption applies.

Preliminary and non-public GPS records may be shared or exchanged with other Federal Government entities by utilizing Memorandum of Understanding (MOU) procedures. The Division of Cadastral Survey is currently studying the possibility of establishing MOU's with agencies interested in obtaining non-public GPS records. Agencies that have expressed interest in obtaining non-public GPS records from cadastral survey offices are the Geological Survey, the Forest Service, and the National Geodetic Survey.

GPS Training

In order to meet BLM's GPS training needs, three distinct training courses have been identified and are currently being offered or are in the development stage. They are, "Advanced GPS", "GPS for Surveyors", and "GPS for Managers". A fourth course in "Resource GPS Applications" is currently being studied.

"Advanced GPS" has been given for the past two years by National Geodetic Survey personnel under a Memorandum of Agreement between National Oceanic and Atmospheric Administration, National Ocean Service and BLM. This course is rich in theory and assumes that the student has had previous experience with satellite survey systems (Doppler) and advanced courses in geodetic surveying or geodesy. "GPS for Surveyors" and "GPS for Managers" are currently under development. "GPS for Surveyors" is being tailored for BLM cadastral survey personnel on the basics of GPS and its application to cadastral surveys; it will be an extension of training usually available to the general public from equipment suppliers, equipment manufacturers, and companies specializing in GPS training programs. "GPS for Managers" is being tailored for BLM management personnel who are outside the Cadastral Survey Program and who are contemplating GPS for their respective disciplines.

CONCLUSION

The BLM has been integrating GPS technology into its Cadastral Survey Program for the past three years. The ease of use and efficiency of GPS have supplanted more traditional forms of establishing geographic coordinates on selected corners of the PLSS.

Inherent to any major infusion of technology is the reality of modifying the existing infrastructure to reflect the needs of the new system. To date, the Cadastral Survey Training Program has been modified to reflect the training needs of cadastral survey personnel, GPS standards for CCS have been establish, and standards for cadastral surveys and resurveys are under way.

REFERENCES

Bureau of Land Management, 1947, "Manual of Instructions for the Survey of the Public Lands of the United States, 1947", Bureau of Land Management, Washington, D.C., 613 pages.

Bureau of Land Management, 1973, "Manual of Instructions for the Survey of the Public Lands of the United States, 1973", Bureau of Land Management, Washington, D.C., 333 pages.

Bureau of Land Management, 1990, "Instruction Memorandum 90-457: Policy for Managing External Access to BLM Records", Bureau of Land Management, Washington, D.C., 28 pages.

Bureau of Land Management, 1990, "Instruction Memorandum 90-567: Policy on the use of Global Positioning System", Bureau of Land Management, Washington, D.C., 16 pages.

Fiedler, J., 1986, "Doppler Handbook", Bureau of Land Management, Denver Service Center, Denver, Colorado, 69 pages.

Fiedler, J., 1989, "Satellite Surveying in the BLM", Presentation at the ASCE Spring Convention 1989, Denver, Colorado, 7 pages.

Fiedler, J., 1989, "Standards and Specifications for Cadastral Control Surveys (Conventional Instrumentation)", Bureau of Land Management, Denver Service Center, Denver, Colorado, 21 pages.

General Land Office, 1930, "Manual of Instructions for the Survey of the Public Lands of the United States, 1930", General Land Office, Washington, D.C., 530 pages.

Honda, J., 1990, "Preliminary Report: Performing Cadastral Surveys with GPS", Unpublished Bureau of Land Management Report, California State Office, Sacramento, California, 7 pages.

White, C.A., undated, "A History of the Rectangular Survey System", Bureau of Land Management, Washington, D.C., 774 pages.

1791 DISTRICT OF COLUMBIA BOUNDARY SURVEY

Michael G. Shackelford, RLS
Chairman
District of Columbia Boundary Bicentennial Committee
P.O. Box 9300, Silver Spring, MD 20906

BIOGRAPHICAL SKETCH

Michael G. Shackelford is registered as a professional land surveyor in five states including Maryland and Virginia, and works in private practice. He is an active member of the American Congress on Surveying and Mapping, the National Society of Professional Surveyors, the American Association for Geodetic Surveying, the Surveyors Historical Society, and is the national chairman for the District of Columbia Boundary Bicentennial Committee.

The Federal Territory Established

With the dissolution of the old Confederation government and the framing of the Constitution of the United States in 1787, it became necessary to find a permanent site for the new capital. Prior to this time Congress had been meeting in various temporary locations including Philadelphia, PA; Annapolis, MD; Princeton, NJ; and New York, NY. Congress began debating on where to locate a permanent seat in 1783, and had suggested sites along the Delaware River near Trenton, NJ, and on the banks of the Potomac River.

The Residence Act, signed into law in July of 1790, authorized the establishment of a permanent seat for the government of the United States of America on the Potomac River. Both Maryland and Virginia had agreed to cede land to the federal government for this purpose. Under the terms of the act, final selection of the ten-mile square that was to comprise the actual capital would rest with President George Washington. Washington's background in surveying and his familiarity with the area surrounding Mount Vernon, Virginia, no doubt influenced his decision on where locate the nation's capital. Both Washington and then-Secretary of State Thomas Jefferson had made surveys of the Potomac River.

Under the Act of 1790, Washington was also authorized to appoint three commissioners to supervise a survey of the site and purchase land for the government. The President selected General Thomas Johnson and the Honorable Daniel Carroll, both of Maryland, and Dr. David Stuart of Virginia (Washington's family physician).

The commissioners were responsible for securing suitable buildings to house the President, Congress, and public offices of the government of the United States by December 1800. They were also to supervise the surveying and planning of the city, which entailed designating and creating city blocks and avenues, selling public lots, and building public structures.

On January 24, 1791, Washington issued a proclamation directing the commissioners to survey the ten-mile square Federal Territory on both sides of the Potomac River so as to include the prosperous village of Georgetown in Maryland and to extend to the Eastern Branch (known as the Anacostia River) of the Potomac River. Because Washington desired that the new capital be situated near good maritime resources, he saw to it that the bustling port village of Alexandria in Virginia was included within the boundaries of the new territory. On March 3, 1791, Congress amended the Residence Act in order to comply with the President's request. It was further provided that no public buildings were to be erected on the Virginia side of the Potomac River.

The Appointment

Major Andrew Ellicott of Philadelphia, Pennsylvania, was selected to perform the survey based on his extensive surveying background, which included several important survey projects in Pennsylvania and New York (locating Pennsylvania's western and northern boundaries, and surveying the entire length of the Niagara River). Additionally, in his possession was one of the most sophisticated collections of surveying instruments that existed in the United States at that time. Secretary of State Thomas Jefferson sent detailed instructions to Ellicott early in February of 1791, and requested that he begin the survey of the ten-mile square at once.

Ellicott accepted the appointment and immediately began his search for an assistant to make the necessary astronomical observations. His brothers, Joseph and Benjamin, had assisted him in the past, but were involved in other surveying projects in New York at the time and were unable to join him in this new commission. However, they did later join him in working on the survey of the Capital and contributed significantly to the completion of the project.

Major Andrew Ellicott's younger cousin, George Ellicott of Ellicott Mills in Maryland, suggested his friend and neighbor Benjamin Banneker be chosen to assist in the survey. Banneker, a free black man who possessed a lifelong interest in mathematics, had succeeded in teaching himself the basic principles of astronomy. The Major was familiar with his qualifications, having reviewed a manuscript for an ephemeris that Banneker was attempting to have published in an almanac. Banneker eventually succeeded in getting his first ephemerides published in a 1792 almanac, and continued to have them printed annually through 1796.

Ellicott's Camp

A base camp for the surveying operations was established near the initial point or apex of the proposed square. This location, known as Jones' Point, was formed by the convergence of Hunting Creek with the Potomac River just south of Alexandria, Virginia.

"Ellicott preferred to establish his main encampment on the top of the highest available elevation in the region to be surveyed, and he customarily sought the protection of the trees or the edge of a forest for additional protection when possible. The focal point of his operation was the observatory tent, which Ellicott located by tracing a meridian and then laying off an angle from it. It was at this observation point that he set up his large zenith sector and near which he placed his astronomical clock. The clock was a critical factor to all his observations,..... he usually set the clock upon the stump of a tree which he had cut down for that purpose. He then erected his observatory tent over the sector and the clock and his other instruments. Other tents for sleeping and for meals were then set up in the vicinity, and an area was provided nearby for tethering the horses." (Bedini 1972).

Major Ellicott, accustomed to commanding large surveying expeditions, was only able to hire six untrained workers for the survey party. The men were required to cut a route through the dense underbrush and forest, and also load and unload supplies from the pack horses. Ellicott traditionally stayed in the camp to perform calculations while competent assistants would perform the field work. However, due to a shortage of field crew members, this survey required that he work in the field while his new assistant, Banneker, who was nearly 60 years old and unable to assist Ellicott in the field on a daily basis, remained at the camp performing calculations, computations, and astronomic observations.

The Instruments

Ellicott's large zenith sector (probably the most accurate scientific instrument in America at that time) was one of his most important surveying instruments. It was used for determination of the latitude by observing stars near the zenith. The six-foot zenith sector was made by David Rittenhouse, a prolific maker of scientific instruments from Philadelphia, PA, and customized by Andrew Ellicott. To observe stars near the zenith, the surveyor, while lying on the ground on his back, had to look through the eyepiece at the bottom of the instrument. This procedure enabled the surveyor to determine the parallels of latitude. Ellicott also had in his possession a small portable zenith sector, a transit and equal altitude instrument, several sextants, three telescopes, two thermometers, two stopwatches with second hands, two sets of cased drafting instruments, two copper lanterns which had special slits for tracing meridians and giving the directions of the lines when they were determined at night by means of celestial observation, and two 2-pole (33 feet) chains.

For taking horizontal observations, Ellicott used a brass circumferentor with a radius of eight inches. He also had a plain surveying compass which was custom made for him by Benjamin Rittenhouse (David's brother) of Philadelphia, PA. This particular instrument was utilized in running the lines and taking bearings between survey stations in areas of dense underbrush. Ellicott fabricated some of the equipment himself, and aside from those items obtained from the Rittenhouse brothers, he acquired the balance of his equipment from the world's most famous makers of optical instruments. One of two achromatic telescopes used by Andrew Ellicott in the field was made by William and Samuel Jones of London, England. This instrument was used for taking signals and observing Jupiter's moons to determine the longitude.

One of Banneker's chief tasks was to maintain the observatory clock, which meant keeping it wound and ensuring it ran at a constant rate. Banneker had to constantly check the temperature of the air surrounding the clock to be sure that the clock's mechanisms were not affected by the cold weather. Even subtle vibrations could affect the clock's precision. Although nights and evenings were the ideal times for taking observations, some of them were performed during the day. In order to establish the correct time for the astronomical clock, Banneker made periodic observations of the sun using the equal altitude instrument.

The Survey

In a letter to Ellicott dated February 2, 1791, Jefferson directed Ellicott to run the first two lines as in the President's proclamation to fix the beginning point, and from that to establish the four "lines of experiment" for the ten-mile square. Ellicott was instructed to find the true meridian, determine the latitude, and map the course of the rivers within the segment surveyed. Jefferson also requested that he note the magnetic variations.

Ellicott completed hiring the necessary personnel, purchasing equipment and horses, and finalized all other preparations, but was forced to delay the start of the survey due to inclement weather. On the evening of February 11, 1791, Ellicott was able to make his first astronomical observations. He ran two preliminary lines the next day and quickly discovered that an adjustment to the proposed survey layout would be necessary. Ellicott determined that Alexandria's wharves and harbor would not be included within the new territory, nor would the proposed survey create ten miles square with straight lines. He submitted a plan that suggested slight deviations to correct this situation, which Jefferson subsequently approved.

Ellicott had a simple plan for laying out the ten mile square: "...(he) traced a meridian at Jones' Point on the west side of the Potomac River and then laid off an angle of 45 degrees from this meridian to the northwest, and continued a straight line in that direction for ten miles. He made a right angle at the termination of this line with a straight line which he carried in a northeasterly direction, also for ten miles and then from the termination of this second line he carried yet a third line for the same distance at a right angle to it, to the southeast. Finally he carried a line from the terminal point at Jones' Point to meet the termination of the third line. He measured these lines by means of a chain, which he examined and corrected each day to ensure that the links had not opened and that there was no other change affecting its accuracy. He plumbed it wherever the ground proved to be uneven, and traced it with his transit and equal altitude instrument." (Bedini 1972).

The Arrival of L'Enfant

The preliminary survey was completed by mid-March. Ellicott established an office in Georgetown, where he was joined by Major Pierre Charles L'Enfant. L'Enfant, a French-born military engineer and architect, was selected to prepare detailed drawings of the actual city to be created within the ten-mile square and to plan the location of the buildings to be erected there. It should be also be noted that contrary to popular belief, Ellicott was the first to receive the appointment to perform a survey - L'Enfant's appointment was not made until the following month.

The Hardships

Ellicott rode out to camp each day before daylight and stayed until after dark; then he would return to Georgetown to work on his maps. Extreme spring weather and the hazardous nature of the trade hampered the progress of the project. Ellicott reported a number of his men killed, one by the falling of a tree.

The country at that time was mostly unexplored and unsettled. All supplies had to be transported by horses, canoes, and men. The men were poorly clothed, with only canvas tents for shelter. The crews were often separated from their families for long periods of time. Additionally, they had to contend with unfriendly Indian forces, wild animals, poisonous snakes, and disease-carrying insects. Throughout the duration of the survey, Ellicott and many of his crew suffered from bouts of influenza.

Despite severe pain brought on by influenza, Ellicott continued to work with his men in the field seven days a week to attempt to complete the project as quickly as possible. Ellicott often stayed for weeks and sometimes months on the assignment without leaving to return home to his wife and family in Pennsylvania.

The Dedication of the Initial Point

The initial point or cornerstone was dedicated on Friday, April 15, 1791, during a ceremony held at Jones' Point near Alexandria, Virginia. Townspeople and officials, including the Mayor, turned out to participate in the event. The Master of Masonic Lodge No. 22 and Commissioner Stuart, assisted by some of the other Masonic Brothers, placed the stone after Ellicott had determined the precise point for its installation. A deposit of corn, wine, and oil were then made upon the stone as part of the Masonic ritual.

The temporary stone at Jones' Point was replaced by a more suitable marker on June 21, 1794. This permanent monument set by Thomas Freeman, an assistant surveyor under the orders of the commissioners, was inscribed with the words: "The Beginning of the Territory of Columbia".

The Jones' Point cornerstone was the first of forty stones that were placed a mile apart and numbered clockwise, one to nine, from corner to corner. Fourteen of the original forty boundary stones were located along the lines of the thirty-three square miles that the Commonwealth of Virginia had ceded to the Federal Government by the Act of December 3, 1789. This land was returned to Virginia by the Federal Government in the Retrocession Act of 1846. All of the stones in Virginia were set in 1791 while those set on the Maryland borders were placed in 1792. Of the 38 monuments that can be located today, it is not known how many of them are original stones or still mark the District boundary.

Banneker's Departure

It had never been intended that Banneker would remain with Ellicott for the duration of the survey. Shortly after the April 1791 dedication of the initial stone, Ellicott's brothers, who had been unable to join him earlier, completed their work on the New York survey and came to assist him. Around the same the time, ill health forced Banneker to return to Ellicott Mills, where he continued to work on his ephemerides, after having contributed significantly to the Virginia portion of the survey.

With his experienced brothers and two other well-trained young surveyors, Ellicott was able to make better progress on the survey. Ellicott was being pressured by the commissioners to complete the work as soon as possible so that they could begin selling public lots. He was personally supervising the boundary lines and the work of his brothers, laying out the streets in the city according to the plan devised by L'Enfant.

L'Enfant's Dismissal

The relationship between Major L'Enfant and the commissioners had been deteriorating for some time. The issue of selling public lots within the District boundaries finally came to a head in October of 1791 when L'Enfant refused to release copies of his original Federal City map for this purpose, after 10,000 copies of the map he had ordered from a French printer never arrived.

Major L'Enfant later insisted that a house under construction by Commissioner Carroll's nephew, Daniel Carroll of Dudington, be removed from the path of a proposed main avenue (now known as New Jersey Avenue). "Since Mr. Carroll did not comply with this demand, the fiery Frenchman gathered some of the workmen and saw to the job of demolition himself." (Grant 1932).

L'Enfant was discharged from the project by Washington, and in March 1792, Thomas Jefferson wrote to notify the commissioners that L'Enfant's services were had been terminated. Ellicott was placed in charge of the entire project, including the plan of the city. He was to proceed with the laying out of avenues and streets in accordance with L'Enfant's original plan.

This was a difficult time for Ellicott as he and his brothers, as well as some of the other workers, were ill with influenza. The Major was now being called upon to work extremely long hours in order to fulfill his responsibilities. He continued to be hounded by the commissioners, who failed to understand his meticulous surveying methods and his adherence to detail and accuracy. Knowing that poor surveying techniques would create future problems in the delineation of the lots within the city, Ellicott refused to compromise his methods.

Ellicott's troubles with the commissioners took him to the point of giving notice before the survey of the city was completed. The commissioners terminated his appointment; however Washington and Jefferson, who held him in high regard, urged him to resume his work on the survey. Washington stated that if it had not been for Ellicott, no plans would have resulted for the capital city.

According to legend, Ellicott reconstructed L'Enfant's plan in detail based on Benjamin Banneker's recollection. However, Ellicott actually used his own sketches and notes made while working with the Frenchman to complete the design with the help of his brother Benjamin Ellicott, who had assisted L'Enfant in drafting his plan. Banneker had left the survey shortly after L'Enfant arrived, and there was little chance the astronomer would have had time to memorize the plans which the Frenchman would hardly have begun.

The Final Map and Certificate

A signed certificate recorded with the commissioners, dated January 1, 1793 states: "...These lines are opened, and cleared forty feet wide, that is twenty feet on each side of the lines limiting the territory: And in order to perpetuate the work, I have set up square milestones, marked progressively with the number of miles from the beginning on Jones' Point, to the west corner, and thence from the west corner, to the north corner, thence from the north corner, to the east corner, and from thence to the place of beginning on Jones' Point; except in a few cases where the miles terminated on declivities, or in waters; the stones are then placed on the first firm ground, and their true distances in miles and poles marked on them. On the sides of the stones facing the Territory is inscribed, 'Jurisdiction of the United States', on the opposite sides of those placed in the commonwealth of Virginia, is inscribed 'Virginia', and of those in the State of Maryland, is inscribed, 'Maryland'. On the third, and fourth sides, or faces, is inscribed the Year in which the stone was set up, and the present variation of the magnetic needle at that place. In addition to the foregoing work, I have completed a Map of the four lines, (with an half mile on each side) including the said District of Territory, with a survey of the different waters....... Witness my hand this first day of January, 1793. Andw. Ellicott". (National Archives 1791-1795).

The United States Coast and Geodetic Survey, now the National Geodetic Survey (NGS), began a resurvey of the District in 1881, which indicated that the original lines laid out in 1791 were slightly in error. For instance, the North point is approximately 116 feet west of the meridian running through the South point. Each of the sides exceed ten miles in length, with the approximate excesses being 230.6 feet on the southwestern side, 63.0 feet on the northwestern side, 263.1 feet on the northeastern side, and 70.1 feet on the southeastern side. Despite these "irregular" measurements, they represent a degree of precision remarkable for the surveying instruments with which they were made and the harsh conditions under which they were accomplished.

Subsequent Events

In December 1914, the Daughters of the American Revolution (DAR) Committee on Preservation of Historic Spots and Records for the District of Columbia held a meeting to discuss the condition of the D.C. boundary markers and explain the need for their preservation. On April 7, 1915, the Committee passed a resolution that the D.C. DAR chapter take up as part of their patriotic work for the year, the preservation and protection of those boundary stones. This work included placing an iron fence or cage around each stone monument - each of the sponsoring chapters paid one dollar for a deed which gave permission to erect the fences. The DAR made arrangements to construct and install the iron fences with the corner posts set in concrete at a cost of $18 each for the intermediate boundary markers. The four corner stones' fences were much larger and cost $43 each to construct. A bronze plaque was placed on each fence bearing the name of the chapter responsible for that stone's maintenance.

The National Capital Planning Commission issued a Bicentennial Report in 1976 entitled "Boundary Markers of the Nation's Capital: A Proposal for Their Preservation & Protection." In this report, the poor condition of the boundary stones was documented and a plan of reommendations was outlined. These recommendations included the establishment of a boundary stone museum at the Jones' Point Lighthouse as well as a maintenance program to preserve the boundary stones and fences.

Future Plans

In 1989, a sub-committee of the ACSM 50th Anniversary Committee was created to begin planning events commemorating the upcoming 1991 Bicentennial of the original District of Columbia survey. It was decided that the National Society of Professional Surveyors should take a lead role in these activities. Subsequently, the District of Columbia Boundary Bicentennial Committee was formed to coordinate the efforts of various other organizations, groups and agencies which are planning similar celebrations.

This joint NSPS/AAGS committee has already begun a Global Positioning System (GPS) resurvey of the Federal territory markers with the help of surveyors from the Potomac Chapter of the Maryland Society of Surveyors and the Mount Vernon Chapter of the Virginia Association of Surveyors. These surveyors come from both public and private sectors, including such organizations as NGS, the Washington Suburban Sanitary Commission, the Fairfax County (VA) Surveyors Office, and a number of private consulting firms.

The technical results of this GPS resurvey will be released in a book to be published by the Committee with assistance from the American Congress on Surveying and Mapping (ACSM). This book will also include a section containing a more complete history of the original survey.

A re-enactment survey and dedication ceremony are planned to take place at the initial or south point of the original survey. The bicentennial celebration, co-sponsored by the Office of Historic Alexandria, is scheduled for Sunday, March 24, 1991 at Jones' Point Park in Alexandria, Virginia. This event will coincide with the American Congress on Surveying and Mapping's 50th Anniversary celebration during the ACSM-ASPRS/Auto Carto 10 Annual Convention and Exposition being held March 24-29, 1991 in nearby Baltimore, MD.

ACKNOWLEDGEMENTS

The author wishes to thank Richard Witmer and Dexter Brinker for their valuable help and ideas. David Doyle, Alan Dragoo, Jeff Lawrence, Doug Richmond, Kim Ripley, Burt Sours, Curt Sumner, and many others contributed significantly to the overall project. Wendy Lathrop helped edit an earlier version of this paper that appeared in the August-September 1990 issue of P.O.B. magazine, Volume 15, Number 6, pages 10-20. Silvio Bedini of the Smithsonian Institution is to be credited for much of the historical material contained in the papers. All of these contributions are gratefully acknowledged.

References:

Bedini, Silvio, 1972. <u>The Life of Benjamin Banneker.</u> Landmark Enterprises; Rancho Cordova, CA.

Bedini, Silvio, 1975. <u>Thinkers and Tinkers, Early American Men of Science.</u> Charles Scribner's Sons; New York, NY.

Bedini, Silvio, 1976. "Andrew Ellicott, Surveyor of the Wilderness". <u>ACSM Journal of Surveying and Mapping.</u> (June 1976): pp. 113-135.

Bowling, Kenneth, 1988. <u>Creating the Federal City, 1774-1800: Potomac Fever.</u> American Institute of Architects Press; Washington, D.C.

Caemmerer, H. Paul, 1950. <u>The Life of Pierre Charles L'Enfant, Planner of the City Beautiful, the City of Washington.</u> National Republic Publishing Company; Washington, D.C.

Conley, Kevin, 1989. <u>Benjamin Banneker, Scientist & Mathematician.</u> Chelsea House Publishers; New York, NY and Philadelphia, PA.

Ellicott, Andrew. Correspondence to Sarah Ellicott. Curtis Collection of Andrew Ellicott Papers, Manuscripts Division, Library of Congress; Washington, D.C.

Grant III, Lt. Col. U.S., 1932. "The L'Enfant Plan and its Evolution". <u>Records of the Columbia Historical Society.</u> Vol. 33-34: pp. 1-23. Columbia Historical Society; Washington, D.C.

Junior League of the City of Washington, D.C., 1977. <u>The City of Washington, D.C., An Illustrated History.</u> Alfred Knopf; New York, NY.

Mathews, Catherine Van Cortlandt, 1908. <u>Andrew Ellicott: His Life and Letters.</u> The Grafton Press; New York, NY.

National Capital Planning Commission, 1976. <u>Boundary Markers of the Nation's Capital, A Proposal for Their Preservation & Protection.</u> National Capital Planning Commission; Washington, D.C.

National Capital Planning Commission, 1977. <u>Worthy of the Nation, The History of Planning for the National Capital.</u> Smithsonian Institution Press; Washington, D.C.

National Archives, 1791-1867. Records of the District of the Columbia Commissioners and of the Offices Concerned With Public Buildings. Vol. 1, RG 42. National Archives; Washington, D.C.

Robinson, June, 1989. "The Arlington Boundary Stones". <u>Arlington Historical Magazine.</u> (October 1989): pp. 5-19. Arlington, VA.

A NEW METHOD FOR MATCHING DIGITIZED CADASTRAL MAPS

B. Shmutter & Y. Doytsher
Technion - Israel Institute of Technology
Haifa , Israel

ABSTRACT

Errors inherent in the digitizing process lead inevitably to variations in the locations of boundaries of neighboring cadastral blocks. Hence it becomes necessary to adjust the digitized data in order to compensate for discrepancies and discontinuities in the digitized cadastral information.

For that purpose it is proposed to adopt a procedure commonly used in aerial triangulation. Each map is regarded as an equivalent to a single photograph and a set of neighboring maps as a photogrammetric block. A system of appropriate equations is formed to adjust the "block". Points lying on boundaries of adjacent cadastral blocks are the substitute for the transfer and tie points used in photogrammetry, and traverse or trigpoints available on the maps constitute the control data.

The results of that procedure are: "orientation" data for each map to transform its content to the state plane coordinate system and adjusted coordinates of points positioned on boundaries shared by neighboring maps.

INTRODUCTION

Converting cadastral maps into digital data is usually effected by digitizing the maps one by one and transforming each map separately to the state plane reference system. That procedure causes by necessity an inconsistent digital representation of the maps.

There are several reasons for the discrepancies:

- Errors inherent in the digitizing process.
- Inaccuracies in the drawings.
- Boundaries of the cadastral block shared by adjacent maps assume different digital descriptions.
- One and the same boundary line may be represented on neighboring maps by unequal numbers of points.
- Neighboring maps are not necessarily at the same scale, hence, the transformed coordinates related to two such maps differ with regard to their accuracy.
- There is no uniform distribution of control data on the maps, which again affects the accuracy of the transformed data.

In order to account for the disagreements in the transformed data and to compensate for the different accuracy levels it is proposed to process a set of neighboring maps as a whole, incorporating into the transformation all the points appearing on two or more maps, and utilizing control data available on all the maps. That enables to consider the accuracy of the control data and to allow for the various scales of the maps. As a result of the simultaneous transformation of all the maps constituting the set each feature assumes an unambiguous digital representation.

OUTLINE OF THE METHOD

Consider a schematic case of a set comprising four maps (Fig. 1). Each map is described by a polygon, the boundaries of the cadastral block it represents, rather than by a regular rectangle.

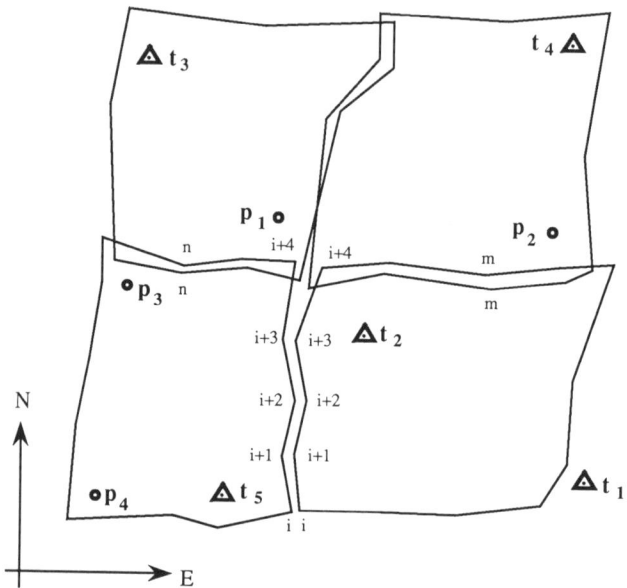

Figure 1. A Set of Maps Prior to the Adjustment

To start the adjustment procedure we need an initial solution, a transformation of each map to the plane state coordinate system. The outcome of such transformations is depicted in the above figure. Usually, such transformations yield contradictory results. For example, the position of the string composed of the points i, i+1, i+2, i+3, i+4 as obtained from transforming the map nr.1 differs from the location of the same string furnished by the transformation of map nr. 2. (The disagreements between the locations of the points are exaggerated for graphical reasons). To compensate for such discrepancies it is necessary to determine for each map transformation parameters ("orientation data"), that when applied to the maps will eliminate the disagreements and define each point unequivocally by a single pair of adjusted coordinates.

Let X,Y be the final coordinates of a point M resulting from the adjustment and x,y the corresponding initial coordinates derived from the map s. The coordinates X,Y are assumed to be related to their counterparts by a linear transformation:

$$X = X_0 + a_1 * x + a_2 * y$$
$$Y = Y_0 + a_3 * x + a_4 * y$$
(1)

Introducing corrections to the "observed" coordinates x,y and assigning approximate values to the transformation parameters yields:

$$X = X + a_1*vx + a_2*vy + dX_0 + x*da_1 + y*da_2$$
$$Y = Y + a_3*vx + a_4*vy + dY_0 + x*da_3 + y*da_4 \quad (2)$$

or,
$$X = X' + b*v + A*\delta \quad (3)$$

X' denotes transformed coordinates obtained from the approximate values of the parameters, v denotes a vector of corrections to the observed quantities x,y and δ a vector of corrections to the parameters.

Now, M not being a control point ought to appear on another map t, otherwise it would not participate in the adjustment. Therefore, another equation similar to (3) can be formulated and referred to that point. But the final coordinates of M must be the same regardless of the map they are derived from, hence:

$$X_s + b_s*v_s + A_s*\delta_s = X_t + b_t*v_t + A_t*\delta_t$$

which provides the following condition equation:

$$|b_t - b_s| \begin{vmatrix} v_t \\ v_s \end{vmatrix} + |A_t - A_s| \begin{vmatrix} \delta_t \\ \delta_s \end{vmatrix} + |X_t - X_s| = 0 \quad (4)$$

Each point common to two maps provides an equation of type (4). Additional equations are derived from the control points (of type (3)).

The set of all common points together with the control points available on the maps furnish a system of condition equations with parameters:

$$B*V + A*Z + W = 0 \quad (5)$$

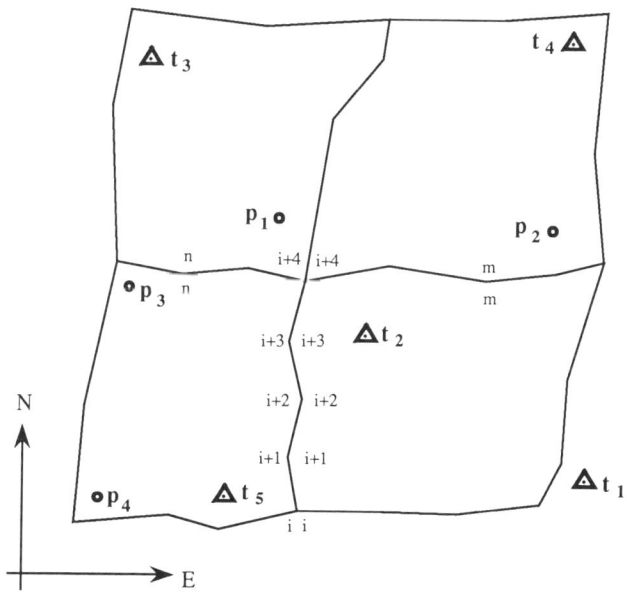

Figure 2. A Set of Maps After the Adjustment

243

The observations forming the condition equations assume different weights according to the nature of the points they are related to. The highest weights are assigned to the control points. Weights equal a unity or less are attached to the other observations, in accordance with the scale of the map they were derived from.

Solving equations (5) by a least squares algorithm provides corrections both to all observed quantities and all transformation parameters. Adding the corrections V to the coordinates and the corrections Z to the parameters completes the adjustment.

The adjusted coordinates of the common points define a consistent layout of the boundaries of the cadastral blocks represented by a set of neighboring maps (Fig. 2).

The final parameters are utilized for transforming the contents of the maps to the state plane coordinate system.

AN ITERATIVE SOLUTION

On many occasions, particularly when there are large numbers of common points, it is advisable to adopt an iterative method for adjusting a set of maps.

The adjustment is effected by several iteration cycles each one consisting of two stages:

a) Computing weighted means for the coordinates of each point positioned on two or more maps.

b) Transforming each map separately to the state plane reference system.

Having executed the first stage a system of observation equations is formed for each map according to expressions (1):

$$X + V = X_a + A*Z \qquad (6)$$

where X_a are the transformed coordinates obtained from the preceding iteration cycle, X are the coordinates resulting from stage a, or coordinates of given control points and Z is a vector of corrections to the current transformation parameters. Obviously, in this case too we discriminate between control and common points by assigning appropriate weights to the observation equations.

Solving equations (6) by a least squares algorithm provides corrections to the transformation parameters and transformed coordinates of points related to the map. The latter are computed according to:

$$X = X_{or} + Z(1) + Z(3)*x + Z(4)*y \qquad (7)$$
$$Y = Y_{or} + Z(2) + Z(5)*x + Z(6)*y$$

Having transformed all the maps (completing stage (b)), stage (a) is executed again and new average locations of the common points are formed. The entire process is reiterated until a predefined convergence criterion is satisfied.

As stated previously, the initial data are already oriented and scaled approximately, hence the alterations of the coordinates arising from the first iteration cycle are within a range of a few meters, and diminish during the reiteration of the process. Consequently, the A matrices and the inverses of the normal matrices associated with each map, as computed during the first cycle, can be regarded as composed of elemnts which do not vary, and need not be computed for iteration cycle anew. Therefore only the vectors of the free terms have to be updated during each cycle. That leads to a fast and efficient adjustment process.

An example of applying the above iteration procedure taken from practice is given below (Fig. 3). Six maps comprising 129 common points and 31 control points (mostly traverse points) participated in the adjustment. Five of the maps were at the scale of 1:1250 and one at the scale of 1:2500. Ten iteration cycles were required to complete the adjustment. A sample of the results is given in table 1.

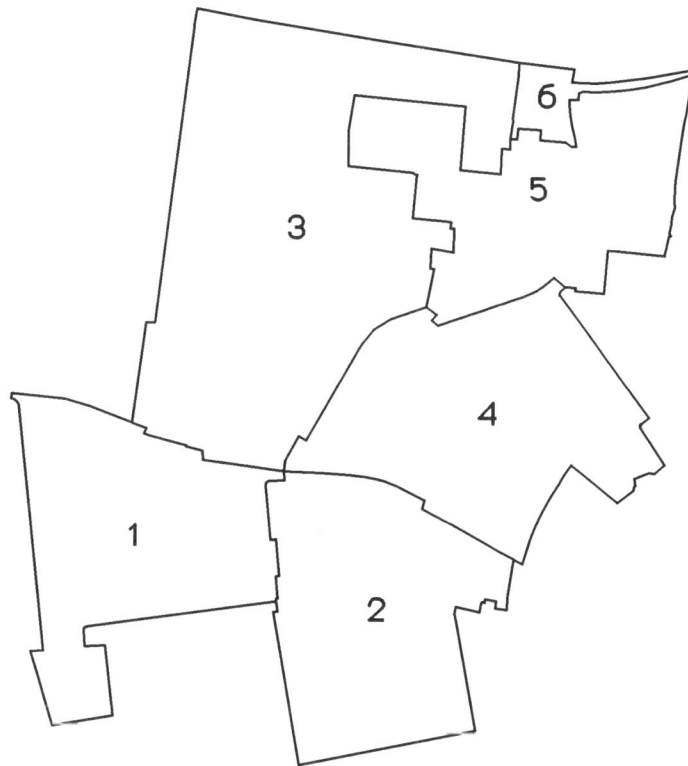

Figure 3. Example Taken from Practice

Two rows appear under the heading transformation parameters, one relates to the corrections derived from the last iteration cycle, the other represents values of the parameters which have been accumulated during the entire process. The first two elements in each row are the translations - corrections to the origin of the map, the remaining are rotation elements combined with scale factors. The standard deviation of an adjusted coordinate computed from the residuals amounts to 0.26 meters, which is equivalent to 0.2 mm at the scale of the map (1:1250).

It is worth noting that the standard deviation of the transformed coordinates computed from all six maps was 0.15 meters and the largest discrepancy between the initial coordinates of a point common to two maps was 2.39 meters, about 2 mm at the scale of the map.

iteration 10   map 5
              map origin   x = 79654.86   y = 57102.36

transformation parameters

.002	.001	-.0000002	-.0000030	-.0000008	.0000073
.339	.348	-.0001332	-.0010151	-.0002845	-.0011349

residuals

vx	vy	vx	vy	vx	vy	vx	vy
.03	-.40	.08	.03	.03	.15	-.16	-.25
-.14	-.06	.38	.06	.19	.00	.09	.02
-.07	.08	-.26	.15	-.10	-.08	-.01	-.07
-.04	-.20	.11	-.15	.13	.37	.12	-.02
.21	-.31	-.37	-.48	-.30	-.06	.30	.02
-.36	-.01	.01	-.08	-.01	.38	-.19	.12
.19	.04	.11	-.03	-.17	-.31	-.08	-.20
.10	-.10	-.06	-.08	.17	.24	-.52	.17
.40	.16	-.30	.30	-.29	.13	.34	-.26
.68	-.46	.57	-.14	.06	.13	.06	.17
-.65	.41	-.33	.24	-.25	.11	.19	.62
-.20	.29	-.44	.11	.38	.02	-.05	-.68
.01	.08	-.19	-.15	.19	-.27	.06	-.15
.03	.14	-.26	.20	.00	.00		

Table 1. Excerpt from the Adjustment Results

SUMMARY

Introducing the concept of block triangulation into the processing of digitized cadastral maps enables to attain a consisted digital description of a set of adjoining maps.

Two adjustment procedures are treated of, a simultaneous and an iterative one. In most cases the iterative method is to be preferred. Its application is straightforward and it is suitable for handling large volumes of data.

A case taken from practice exemplifies the application of the method and its merits.

BACK FROM THE BRINK

Ferret Habitat Survey
at
Wind Cave National Park

Dennis D. Shreves, RLS
Associate Professor of Civil Engineering Technology
Kansas College of Technology
Salina, KS 67401

ABSTRACT

Until the early 1980's the black-footed ferret, a native of the North American prairies, was regarded by many as extinct. In September of 1981 a few of these animals were found in a remote locale in northwestern Wyoming. After several years of close study a distemper plague ravaged that colony. The few surviving animals were trapped, and a breeding program begun. The only known living ferrets are now being held in captivity.

Prairie dogs are the primary food source of the black-footed ferret. Efforts have recently been initiated by the U.S. Fish and Wildlife service to locate areas of potential ferret habitat within the boundaries of public lands. Studies are under way by the National Park Service to determine if it might be feasible to release ferrets into existing prairie dog towns located within the confines of several of their parks. In the summer of 1990 a survey was performed to determine the location, size and density of the prairie dog towns found within the borders of Wind Cave National Park in South Dakota. The results indicate potential ferret habitat may be available in this park.

THE PLIGHT OF THE BLACK-FOOTED FERRET

Ten years ago, the black-footed ferret, a native of the North American prairies, was considered extinct by many wildlife specialists. Then in September of 1981 a dog living on a ranch near Meeteese, Wyoming, killed a strange animal near its food bowl. When it was positively identified as a black-footed ferret Wildlife Biologists were ecstatic.

A member of the weasel family, black-footed ferrets live in prairie dog towns. They will eat mice, rabbits, ground squirrels and other small animals, but estimates are that up to 90 percent of their natural diet consists of prairie dogs. In one night a ferret might visit up to 250 burrows within a range of 150 acres. They also rely on the rodents for shelter, as they live in burrows located within the towns. Ferrets are secretive and nocturnal, traits that may afford them protection from eagles, hawks, coyotes and badgers.

John James Audubon and John Bachman were the first to record the existence of the black-footed ferret back in 1851. A skin was brought to them at Fort Laramie. The original range of the animal is thought to have extended from Alberta to Texas and from Utah to Nebraska (an area roughly defined as the western Great Plains). This area also coincides with the former range of the prairie dog, an animal estimated to have once numbered more than five billion.

The decline of the black-footed ferret is undoubtedly attributable to the elimination of prairie dogs by settlers who considered the rodents pests. They are still viewed by many ranchers as a threat to cattle and crops. Eradication efforts have included the use of cyanide, strychnine, and other poisons. Gas, dynamite and shooting have also been used. Plowing under the towns is another effective control technique. In some states prairie dog populations have been reduced by as much as 99 percent.

Extensive eradication campaigns had never been pursued in the area around Meeteese. Ranchers there seem to have a live-and-let-live spirit.

EFFORTS TO PROTECT THE FERRET

Prior to the Meeteese discovery a small colony of black-footed ferrets had been located in South Dakota in 1971. The U.S. Fish and Wildlife Service collected these animals and instituted a captive breeding program, but for many reasons it failed. By 1977 all of the animals had died and at this point many scientists assumed that the species was indeed extinct.

The Wyoming Game and Fish Department took the early lead role in ferret management for the Meeteese colony. By 1984 at least 128 animals were known to be living in the area. Then nature dealt the animal two serious blows. In September of that year, plague swept through the towns. While the ferret is relatively immune to the disease, it ravaged the prairie dog population. Despite control efforts by wildlife officials the wild ferret population was depleted by more than fifty percent. By October, only 31 animals still survived.

In 1985 as the plague began to dissipate, an epidemic of canine distemper struck the ferrets themselves. This disease could have been transmitted to the colony by a coyote, badger, fox, domestic dog or even a member of the crew dusting the town for the plague. The remaining ferrets were all but wiped out. A census taken in the fall of 1986 indicated that only 21 still survived, and six of these were in captivity. Three of the wild animals were males and two were females with litters of five each. Scientists felt that the animal might have reached a "genetic bottleneck"; that inbreeding now threatened their ability to maintain a viable population.

In September, the Wyoming agency decided to capture the remaining wild ferrets, and institute a captive-breeding program. By November, only six of the wild animals had been collected. The fact that they were probably closely related led to further concerns about inbreeding. These animals were taken to the Sybille Wildlife Research Unit, about 40 miles northeast of Laramie. In the meantime, the six ferrets already in captivity died of distemper.

At the center the animals were contained in a highly protected environment. Strict decontamination procedures are practiced before entering the ferret enclosure. These include a complete scrub down, change of clothing and an antiseptic hand rinse. Pneumonia and influenza are only a few of the human carried diseases that the animal appears to be susceptible to.

These recent efforts have been more successful. By the Fall of 1990, the captive population had grown to 185 animals. Most of these remain at the Sybille Unit, but three have been transferred to the National Zoo at Front Royal, Virginia. Wildlife officials have indicated that additional transfers to other breeding centers will be made, partly to prevent the possibility of disease from destroying the entire population if they are maintained at one site. Geneticists claim that 500 individuals are needed for the species to recover to a healthy status.

In the meantime, the U.S. Fish and Wildlife Service has initiated a search for potential release sites. Habitat has been studied at various locales within their former range, but ideal locations are rare. Public lands are preferred, for the obvious reason, that they can be managed by government agencies. Ideally, the sites should have minimal human interference, be relatively disease free and have adequate numbers of prairie dogs. Several locations appear to meet these requirements. One is the original Meeteese site and another is located near Shirley Basin, Wyoming. Two are in National Parks located in western South Dakota. The most promising of these is in the Conata Basin area of the Badlands. The other is Wind Cave.

WIND CAVE SURVEY

Wind Cave National Park is located on the southwestern slopes of the Black Hills near Hot Springs, South Dakota. This 44 square mile park contains abundant prairie land and more importantly, a thriving population of prairie dogs. A Federal survey team performed a preliminary study in August of 1989, and it quickly became evident that not enough was known about the towns in the park to determine whether or not it would be a good relocation site.

Park officials made plans over the winter months to survey the towns and attempt to determine their size and current populations.

FIELD WORK

Actual field work began in June of 1990. Basically, there were two parts to the project.

Perimeter, Area & Location Surveys

Twenty towns were known to exist at one time or another within the park boundaries. Efforts to determine their locations and sizes had not been made since 1982. The author is not certain of the technique used, but believes the perimeters were measured with compass and perhaps a chain. Distances were more likely determined by pacing. An earlier survey was completed in 1938.

Aerial photographs were used to determine acreage of each town at the Badlands, and it was initially hoped that the same technique could be applied at Wind Cave. However, the park had not been flown since 1980 and it was felt that these photos were probably too old to indicate accurate boundaries. The cost of reflying the park was considered prohibitive.

Therefore, traditional survey methods were used to locate and size each town. A total station was not available for the project, but a digital theodolite (Lietz DT5) and electronic distance meter (Lietz Red2A) were furnished by the Kansas College of Technology. They proved to be adequate, albeit not as efficient as a total station would have been.

The boundary of each individual town was established by setting up on a station in or near each town and having assistants walk the perimeter while carrying targets. Shots were taken about every 50 to 100 meters, and coordinates recorded for each. Standard survey programs (Survey3 and HP41 CX with a Survey Module) were used to determine the area, shape and relative location of each town. Each survey was tied to one or more control points or distinct landmarks within the park. Maps of the towns were generated with AutoCAD such that they could overlay existing quadrangles.

The larger towns were typically shot with one or two setups. Many of the smaller could be completed with one. Several problems were encountered. Getting equipment to remote locations was sometimes a difficult task, especially as the grass dried and the danger of fire in the back part of the park grew. Wild animals, bison in particular, occasionally interrupted the progress of the field work. Rattlesnakes were commonly discovered around the burrows, but outside of the curiosity they always arouse, caused only minor delays. Most of the field work was performed very early in the morning to avoid the afternoon heat.

Density Surveys

The technique used to determine the population density of each town was a little more unique. It was developed by the U.S. Fish and Wildlife Service, the Wyoming Game and Fish Department and the Montana Department of Fish, Wildlife and Parks.

Strip transects were made through each town. Each transect was 1000 meters long by 3 meters in width and therefore included 0.3 hectares. A metric Rolatape wheel was used to determine distance and a 3-meter piece of electrical conduit was attached to the frame of the wheel to establish the width. A small chain braised to each tip of the conduit was allowed to hang freely to within a few inches of the ground. As the wheel was pushed forward, the chains helped facilitate the occasional decision of whether or not to include borderline burrows in the survey. The burrow was counted if more than half of its opening fell within the transect.

Separate counts were kept of the active and inactive burrows within each transect. Active burrows were defined as those that had fresh scat within half a meter of the opening. Occasionally the decision to call the scat fresh was rather arbitrary, but the technique seemed to offer a fair indication of prairie dog activity. One researcher indicated that upon arising in the morning, the first activity performed by a male prairie dog is to investigate the burrows surrounding his own. This could, and probably did, lead to some inactive burrows being counted. At other burrows, where dogs were actually seen, no scat was found at all. Consistency in technique was probably more important than actual activity, and the count of active burrows seemed reasonably fair. To be counted the burrow opening had to be at least seven centimeters in diameter and deep enough so that the end could not be seen.

It was felt that a five percent sample would probably provide a confident estimate of the population for each town. A systematic sampling technique was used. The space between transects could be determined by the following equation.

Spacing = Transect Width / Desired Proportion

To obtain a 5% sample a width of 3 meters divided by 0.05 yields a spacing of 60 meters. We found that a value of 50 meters seemed to insure better coverage. For end-to-end transects a gap of 50 meters was left between each.

The actual operation involved commencing at one end of the town and walking back and forth across it pushing the Rolatape and counting active burrows that fell under the 3-meter conduit. The course was determined, and held, by magnetic compass and it was reversed each time the edge of the town was reached. With a minimum crew of two, about ten to fifteen transects were completed each day.

DATA EVALUATION

Using the area calculated for each town the number of transects to be run to insure a minimum five percent coverage was determined by multiplying the area by 0.05 and dividing by 0.3 (the area of one transect). Southeast Town for instance had an area of 79.25 hectares. The calculation for needed transects thus became:

$$(79.25 \times 0.05) / 0.3 = 13.21,$$

and thirteen transects were run on this town.

Total active burrows for each town were tabulated from the field data. If a transect had twelve or fewer burrows it was not included in the count. This number was used to estimate the percentage of good prairie dog habitat in each town. For instance, if two transects in the Southeast Town had happened to have totals of 12 or fewer active burrows then 11/13 or 85% of the habitat would be considered good. This was not a common occurrence. All of the transects in Southeast Town were good.

There were 432 active burrows in the thirteen transects ran through Southeast Town.

432 Burrows / 13 Transects = 33.23 Burrows/Transect. And,

33.23 Burrows per Transect / 0.3 Hectares per Transect = 110.77 Burrows per Hectare.

For white tailed prairie dogs the burrow density (PD DEN) was calculated by use of the following equation provided by the U.S. Fish and Wildlife Service:

PD DEN = 2.508+(0.150 X Active Burrow Density)/0.566.

For the Southeast Town this worked out to:

PD DEN = 2.508+(0.150 X 110.77)/0.566 = 33.79 Prairie Dogs per Hectare, and

79.25 Ha X 33.79 Prairie Dogs per Ha = 2677.9 Prairie Dogs.

The number of ferret family groups that the colony is capable of supporting was estimated by dividing the prairie dog total by 763 (763 prairie dogs per ferret family). If there were fewer than 272.5 prairie dogs, the town was not included in the count. A ferret family is defined as one male, two females and 3.3 pups. The Southeast town could then theoretically support:

2677.9/763 = 3.51 Ferret Families.

WIND CAVE NATIONAL PARK
BLACK-FOOTED FERRET HABITAT SURVEY
1990

TOWN	HECTARES 1990	ACRES 1938/1982/1990			GOOD	P DOGS /HA	TOTAL P DOGS	FERRET FAMILIES
Back Forty	0.00	0	0	0	0%	0.00	0.00	0.00
Bison Flats	281.91	30	645	697	98%	38.06	10514.90	13.78
Boland Ridge	7.10	40	35	18	100%	31.37	222.73	0.00
Elk Mountain	1.57	15	5	4	100%	23.87	37.48	0.00
Goose Pimple Buttes	0.00	30	0	0	0%	0.00	0.00	0.00
Highland Creek	30.74	120	40	76	100%	40.12	1233.29	1.62
Norbeck	15.76	35	170	39	100%	33.29	524.65	0.69
North Town CSP	75.84	?	?	187	88%	34.28	2274.82	2.98
North Town WICA	21.42	30	110	53	88%	34.28	642.49	0.84
Pringle Cutoff	44.24	10	125	109	100%	33.14	1466.11	1.92
Rankin Ridge	4.90	0	10	12	100%	61.85	303.07	0.40
Red Lick	0.00	0	15	0	0%	0.00	0.00	0.00
Red Valley (Homestead)	4.30	105	20	11	100%	35.35	152.01	0.00
Research Reserve	121.02	95	270	299	100%	31.33	3791.56	4.97
Rocky Knob	0.00	0	5	0	0%	0.00	0.00	0.00
Sanctuary	9.04	0	135	22	100%	34.91	315.59	0.41
Shirttail Canyon	0.00	45	35	0	0%	0.00	0.00	0.00
Southeast Town	79.25	0	170	196	100%	33.79	2677.86	3.51
Upper Highland	0.00	145	5	0	0%	0.00	0.00	0.00
Wind Cave Canyon	1.30	0	5	3	100%	22.99	29.89	0.00
TOTALS	698.39	700	1800	1726		34.63	24186	31.12

Ferret Family = 1 Female, 0.5 Male, 3.3 Young

Similar calculations were made for all towns in the Wind Cave complex and the results tabulated with a Lotus 1-2-3 spreadsheet. This table indicates that the park could currently support about 31 ferret families. However, several other considerations need to be made.

THE RESULTS

Is the data relevant? Obviously, until ferrets are released and more studies are done, all of the preceding calculations amount to best guesses. A good deal of research went into the development of the technique and the quantitative analysis, but major questions still remain, and until ferrets are actually turned loose we won't have the answers to most of them. Until that time, we can't know if the quantitative data is valid.

The biggest question probably concerns fluctuations in the size of towns. The limited data available suggests that prairie dog towns in the park are subject to fairly rapid changes in size and density. Some of the causes are obvious. Over the years the Park Service has made efforts to control the size of the towns, with particular attention being paid to those located near the borders. This has been done primarily to prevent migrations of the animals onto neighboring lands. In the early eighties a decision was made to attempt to reduce all of the towns to their historical size as determined by the 1938 survey. Some towns were completely eradicated in this campaign, but many of these seem to be making a comeback. On the other hand, the Shirttail Canyon Town which was not touched, has for no obvious reason completely disappeared. There is speculation that it could have succumbed to the plague or some other disease.

Efforts were made in the summer of 1990 to relocate some of the prairie dogs from a few of the border towns to the interior of the park. This project does not appear to have been overly successful, but obviously, similar projects could have a future effect on the size and density of the involved towns.

Predators that thrive on prairie dogs and ferrets could also have a significant impact on any reintroduction program. Not much is known about ferret predators.

A major U.S. highway runs through the heart of Bison Flats, the largest town in the park. Ferrets are nocturnal animals and therefore may become potential road kills.

Another big concern is the fact that there are currently no known ferrets in the wild. Can ferrets bred and reared in captivity adapt to a natural environment? Estimates of expected mortality, when they are released, run as high as 90%.

THE FIRST RELEASE

The first reintroduction is scheduled for the Fall of 1991. The site will probably be either near Meeteese or Shirley Basin, in Wyoming. If successful the Badlands and Wind Cave are also being considered to receive ferrets in the following years. However, much additional work needs to be done at the latter park.

All of the towns at Wind Cave should be monitored for additional size and density changes. Particular attention needs to be paid to the Research Reserve, Red Valley and Sanctuary Town. These are all in relatively remote locations, and the latter two have potential for quite a bit of growth.

Further efforts and study should perhaps be made on the feasibility of transplanting prairie dogs from border towns to the interior of the park. The efforts made in the summer of 1990 were admittedly haphazard. Might it be possible to support more ferrets in a town by occasionally "importing" more prey?

The loss of a single species on this planet is irrevocable. It is a permanent loss for us all. Surveyors and mappers have a place in the prevention of this type of tragedy, and we all have a responsibility to make certain it does not occur. The black-footed ferret is an endangered species. "Endangered" means there is still time.

REFERENCES

Biggins, D., Miller, B., Oakleaf, B., Farmer, A., Crete, R., and Dood, A., 18 May 1989, A System for Evaluating Black-Footed Ferret Habitat, Report Prepared for the Interstate Coordinating Committee. U.S. Fish and Wildlife Service; Wyoming Game and Fish; Montana Department of Fish, Wildlife and Parks.

"Baby Boom", Natural History, v97, (August 1988), 6.

Clark, Tim W., "Last of the Black-Footed Ferrets?" National Geographic, v163, (June 1983), 828-838.

"The Plight of the Black-Footed Ferret", Discover, v7, (February 1986), 7.

Raloff, Janet, "Captivity Chosen to Save the Ferret", Science News, v130, (Sep 6, 1986), 151.

Richardson, Louise, "On the Track of Black-Footed Ferrets", Natural History, v95, (February 86), 69-77.

Weinberg, David, "Decline and Fall of the Black-Footed Ferrets", Natural History, v95 (February 1986), 62-69.

Williams, Ted, "The Final Ferret Fiasco", Audubon, v88, May 1986), 110-119.

A DISCUSSION OF DUALISM IN PUBLIC SECTOR LAND INFORMATION

Nancy von Meyer
Fairview Industries
8616 Fairway Place
Middleton, Wisconsin 53562

ABSTRACT

In the U.S., governments are often the largest holders of land related information. In terms of land information, public agencies generally take one of two roles, either an information holder and manager or an information collector and gatherer. The role an agency plays in the land is critical to defining the type, extent, completeness, and accuracy of the resulting land information. This paper examines some of the issues related to the dual roles of governments in land information.

INTRODUCTION

Without accurate information about the lands and waters, and without an up-to-date inventory of a country's land resources and what is happening to them and the environment, the government and the people are handicapped in controlling their own destiny. It is not possible to make the best use of the land and natural wealth, or to prevent its mis-use, without good factual knowledge of the country and its features. (Root, 1985).

Land is the foundation of the American culture and economy. Nearly every activity of society impacts the land and depends on it as a base. As land use becomes more intense and complex, competition for specific tracts of land and resources is escalating. Decisions concerning the allocation and use of the land are made routinely by private land owners, investors, developers, and agencies at all levels of government. To a large degree, these decisions are shaped by social, economic, and political values. But, the effectiveness and efficiency of decision-making depends on the information available to the participants. (BSC, 1989)

It is not just a question of the quantity of information about land, but whether the information is accurate, accessible, and appropriate. Determining what information exists, where it is located, who collected it, and whether it can meet current needs is often a major problem in the decision-making process, and hence a major factor in the design of land information systems to meet the decision-maker's needs.

"Public information is gathered, received, processed, managed, or held by a government agency in the course of its statutorily mandated activities." (Epstein, 1990). At issue is the difference between information gathered by a government and information held by a government.

FEDERAL GOVERNMENT ROLES

The federal government has two distinct positions or roles with regard to land information. In one, the federal government acts as an information provider, analyzer, and user. For example, one portion of the Bureau of Land Management (BLM) is a collector and provider of information on federal lands, acting as a large land holder, or custodian. As an owner, on behalf of the American people, the BLM maximizes the use and value of the land for all shareholders, in a caretaker arrangement.

In a second role the federal government acts as a model and resource for local and state projects requiring technical expertise and data standards. As examples, the National Geodetic Survey (NGS) provides a nationally based reference system for mathematically defining positions on the earth and the Federal Information Processing System (FIPS) defines data standards for computer systems and databases. These are broad based, content neutral technical resources.

Determining the type, extent, completeness, and accuracy of federal land data depends in part on which role the federal government played in assembling the information. Likewise the design of information systems to manage the data, depend on what role is served. For example, two database systems for coordinate values in the federal government are: the NGS's data system for the National Geodetic Reference System (NGRS), and the BLM's developing Geographic Coordinate Database (GCDB) for coordinate values on section corners. These databases have similar content, but were developed from different roles. The NGRS is a national resource, which serves as technical guidance and a standard for others to follow. It is provided to users as a basis for other data. The GCDB is the federal government's landholder information on federal lands. It is inherently linked to many other data systems which provide descriptions of the extent and types of land rights. The design and expected output of these two systems are different.

STATE GOVERNMENT ROLES

States also have a dual role. For example, in Wisconsin, the Department of Natural Resources (DNR), representing the State, plays a landowner role in the management of state parks and nature preserves. But, the State also serves as a resource for county and local governments in land information system design and development.

The Wisconsin Land Information Program (the Program) was created in 1990, Wisconsin Act 31. Act 31 also created the Wisconsin Land Information Board (the Board) and gave it responsibility to implement the

Program. As a part of the Program, the Board developed a set of recommendations and guidelines to assist counties with the development of county-wide land information and land records modernization plans. These guidelines provide a common conceptual framework for the counties, but do not define specific implementation constraints.

The design of an information system for the DNR to manage state lands has a very different set of principles than the confederated network of land information systems described by the Board. For example, as a landowner the DNR has interest in identifying every parcel of land they own and requiring all permits and licenses to conform to strictly established and enforced guidelines. On the other hand the Board, a state organization managing a state-wide land information program, provides guidance on state-wide data exchange.

It is the intent of the Board to afford each County the widest possible latitude in its interpretation of land records modernization and the development and implementation of the County-wide Plan. However, this must be balanced by the need for effective use of public funds for programs which are consistent and efficient, and will ultimately contribute to a fully confederated network of land information systems. (WLIB, 1991).

Land information required by state agencies to manage state owned lands is different than the information required to facilitate state-wide data exchange of land information maintained by counties. Both roles are identified as "State roles," but the resulting systems and standards are not identical.

COUNTY AND LOCAL GOVERNMENT ROLES

County and local governments also have dual roles, but the focus of most information system papers and reports focus on the information holder and manger role. In this capacity county and local governments play a key role in building and maintaining comprehensive land information systems for all other levels of government.

Most county and local governments take a passive role in land records in that they are passive receivers of information. Perhaps the most obvious evidence of this is a passive register-of-deeds system which must accept and hold any and all recorded documents. There are very few places in the country where a register-of-deeds can refuse to record a document based on inaccuracies or incompleteness in content.

There are exceptions to and changes in this passive role. As examples, many land subdivision statutes require content and data quality review prior to recording. And, in some jurisdictions subsequent government users of register-of-deeds information, such as tax listers, cadastral mappers, or treasurers, can refuse to service land until the information contained in the deeds is accurate.

HOLDERS AND GATHERERS

Some of the issues associated with land information system design hinge on whether the information system is intended to service passive information holders or active information gatherers. If a system is designed to service collections of information records, it should be designed to address issues of inaccuracy and incompleteness.

For example, a system developed from existing register-of-deeds files may not provide complete coverage of title or accurate definition of the spatial extent of rights in land. The authority of such a system to enforce corrections to historical and potentially incorrect legal documents may not exist. The question of how well the public is or is not served by this system is not issue if the authority to correct or to enforce completeness and accuracy does not exist.

A system designed to service an information gatherer role can provide for standards of completeness and accuracy. For example, foundations and federal agencies, which control large tracts of land, may, within their boundaries, specify measurement procedures, types of records, and completeness of coverage. The system design issues become cost of accuracy and cost of coverage, rather than absence of records.

A land information system design which uses or is based on public sector land records, should consider which role the government played in assembling those records. The government's role will affect the accuracy, completeness, and authority to correct the records. It is also possible that the government's role may have an affect on the access and ownership of the resulting land information.

REFERENCES

BSC Group, (1989), <u>Managing Our Land Resources</u>, Executive Summary, Report to the U.S. Department of Interior, Bureau of Land Management, Washington D.C., May, 20 pages.

Clapp, J.L., (1990), "Approaching LIS as a Public Service Facility," <u>ACSM Bulletin</u>, August, page 43.

Epstein, E., and McLaughlin, J.D., (1990), "A Discussion on Public Information," <u>ACSM Bulletin</u>, October, 1990, page 33.

Root, E.F., (1985), "Surveys and mapping to aid environmental control and management of resources," in <u>Proceedings of the United Nations Inter-Regional Seminar on the Role of Surveying, Mapping, and Charting in Country Development Programming</u>, Aylmer, Quebec; Energy, Mines and Resources, Ottawa Canada.

Wisconsin Land Information Board (WLIB), (1991), <u>Recommendations and Requirements for County-Wide Plans for Land Records Modernization</u>, Wisconsin Department of Administration, Madison, Wisconsin, 19 pages.

DEVELOPMENT
OF AN
ELECTRONIC FIELD BOOK
FOR
CADASTRAL RETRACEMENT SURVEYS

Jerry L. Wahl
Branch of Cadastral Surveys
Bureau of Land Management
California State Office
2800 Cottage Way, E-2841
Sacramento, California 95825

Raymond J. Hintz
University of Maine

Corwyn J. Rodine
Bureau of Land Management
Eastern States Office

ABSTRACT

This paper is a discussion of the background for development of an Electronic Field Book whose requirements for data collection and other features are directly applicable to boundary retracement surveying. This application of data collection offers unique challenges that are not specifically addressed in commercial data collection systems. Among these attributes are high efficiency in traverse as opposed to radial survey, and onboard computation capabilities specific to the U.S. Public Land Survey System. An additional requirement is a thorough and easy to use mechanism for recording descriptive and evidentiary information as well as topographic and planimetric data relevant to the legal aspects of boundary surveying.

BACKGROUND

Since 1988 the Bureau of Land Management has been actively investigating the feasibility of survey data collection for it's Cadastral Survey function. At that time data collection was not generally considered worth implementing on surveys involving any significant amount of traversing. Various studies of data collection in the 1985 time frame as well as common sense indicated that station setup time overrode measurement time savings in all but pure radial shootout applications.

Neglecting the lack of field advantage, it was still recognized that some efficiencies were achievable in post processing with integrated software. In actual practice these advantages were hard to realize because of the lack of an effective common post processing system at the time within Cadastral Survey BLM, and that data collection capable surveying instruments were not generally available or in use.

Other factors which had led to failure of earlier attempts at using data collection were an apparent large *implementation curve*. That is, problems in setting up a system frequently prevented a user from becoming operational. Things as simple as obtaining a proper cable, or learning a different field procedure were common. Investigations indicated that there were a significant percentage of purchasers of data collection systems that ultimately did not use them.

Nevertheless over the years a few individuals within BLM Cadastral Survey had successfully implemented data collection systems via the HP-41C even though the primary advantage gained was automation of self written post processing. No commercial systems were successfully implemented.

Changing Times:

By 1988 things were changing. First, PC's began going along with the field surveyor to remote projects. The first use of the PC was to assist in preparation of the official record field note returns required in BLM Cadastral work. With the PC there was at last the potential for a place for the data to go. This was enhanced by the emerging automation of other portions of the *field to finish* cycle. Computerized plat drafting was becoming established, and in some offices survey computations were being performed within an AutoCad environment. Also at this time the slow conversion and upgrading of field survey equipment had allowed data collection capable EDM and electronic total stations to start being available 'on-line'.

Out of this scenario there arose an increased desire by a number of field people to try data collection. As a result Cadastral Survey began investigating data collection in order to identify what capabilities would be desirable for BLM Cadastral Survey applications, as well as what things seemed to be hindrances to success.

The approach taken was to try a number of different existing systems based upon proposals received for individual projects. Hopefully this would allow for feedback and the individual surveyors making the proposals would follow through and implement the systems allowing for a thorough evaluation.

As a result of these test projects and other field input as an ongoing process, much was learned about desirable attributes of a Cadastral data collection system.

Test Projects - Prototyping

Through a number of these prototype tests Cadastral learned a number of lessons working with the following 5 classes of data collection systems.

1. 41C stand alone systems: These systems were typified by no interface to an instrument, but used real time keyed in measurement data. Advantages were that implementation was basic, inexpensive, available to almost everyone for the cost of a PC interface, flexible, and defined by the user.

    Lessons: These systems have good flexibility, field computations, low size and weight, but lacked capacity and data integrity.

2. 41C systems using the Topcon HA1 interface: Cadastral's Eastern States Office developed a system using this interface. This is one of the most successful functional implementations.

    Lessons: The system functions with only a few Topcon instruments, there was need for more storage, and full geodetically correct field computations that were part of the system were not available on line.

3. Systems based on the Tripod Data Systems (CO-OP41) data collection sub-system: These have the advantage of considerably more storage, built in communication with PC's and many instrument types, and improved data integrity. They can be enhanced with in-house ROM's.

    Lessons: This system contains the desired ability of working with multiple instruments. Limited descriptive fields are due to 41C limitations. Flexibility and customising are still desired and achieved through custom EPROM's.

4. CMT MCII and MCV handheld's: These units fall midway between the 41C systems and the true MS-DOS PC's. They are light-weight with moderate amounts of storage. The 41C language available for the unit allows flexibility and with the optional TSTATION software a custom data collection system can be created in 41 code as well as in C.

    Lessons: The MC II has no environmental integrity, while the MCV has this for a demonstratively larger cost. The rapidly improving hand held PC's have dropped cost wise and offer much more flexibility. If an instrument is not full MS-DOS it can inhibit use of DOS based programs.

5. Dedicated data collectors: Nikon, Lietz SDR24, Zeiss DAC500 were also tested with mixed results. Typically problems were based on proprietary procedures, cost, and lack of on-board geodetic computations.

CURRENT DEVELOPMENT

As a result of the testing and feedback it became apparent that the most likely course of action was to pursue custom development of a data collection system that came to be called the Cadastral Electronic Field Book (CEFB).

During the last few years Cadastral Survey has also been developing a custom PC based Cadastral Computational system known as Cadastral Measurement Management (CMM). This software is in beta test at this time, and was developed through a co-operative effort between the Bureau of Land Management (BLM) and the University of Maine (UM). CMM is expected to be the first iteration in an effective PC-based dependent resurvey computation system. The pursuit of field data collection systems is closely coupled to that software development.

Two of the most difficult problems associated with CMM are the monotony and error prone task of manual entry of field measurements, and the inability to have CMM functionality in a field environment. The CEFB effort is now just beginning, however due to the testing and prototype efforts, it's major components have been identified and software prototypes for each component currently exist.

The development and testing of CEFB will follow the same successful strategies which have been discussed regarding similar factors related to CMM (Rodine, et al., 1991). A significant amount of interaction between BLM and UM will result in a field test of CEFB, and results of the field test will enable modification of CEFB prior to a beta release.

Why another data collection system?

Many would argue that a plethora of data collection systems now exist, and to create another would surely be redundant. This could be true for many applications, but no system exists which truly fulfills the dependent resurvey needs of the U.S. Public Land Survey System (PLSS). If one is to totally replace the field tablet with an electronic equivalent in the Cadastral Survey application the following must be considered:

* Ability to efficiently handle traverse based work, often through difficult terrain. Job procedure may vary widely to adapt to terrain and logistics. The interface to both machine and surveyor cannot dictate a procedure.

* Ability to work with multiple instruments, as financial practicality and federal procurement restrictions will not likely allow various offices to obtain uniform instrumentation.

* Ability to fully and flexibly describe points, evidence, monuments, topo, testimony and other legally significant descriptive data.

* Ability to perform many classes of Cadastral specific computations needed in the field, including astronomic observations, offsets, corner setting, etc.

* Provide legal and practical data integrity by collecting raw data and reducing and verifying data in real time as well as computing other pertinent line position data.

* Collection of geodetic coordinates in addition to raw measurements is a necessity in a dependent resurvey.

* Field note automation can be more successfully implemented when one considers the uniqueness of the PLSS.

* Geodetically correct PLSS computations such as single proportion, double proportion, midpoint, etc. must exist in the field unit.

* The system must accept geodetic positions from a system dedicated to the PLSS such as CMM (Rodine, et al., 1990).

Many other highlights of CEFB will be detailed in this paper which do indeed exist in other data collector systems.

Hardware Platform.

The choice of a data collection hardware was driven by a number of factors. Primary among these are future potential of the system, flexibility, size, weight, cost and, of course, functionality. This could have resulted in a difficult choice, however the criteria developed in the testing, feedback and prototyping phase limited the prospects to only a few options. For example, the requirement to interface to a multitude of instruments is very limiting. The 41C based systems have proved to be the most flexible, but also the most limited in storage or CPU capability and had no future due to HP's discontinuance of the 41C in mid 1988.

The choice was settled on the emerging MS or PC-DOS compatible hand held PC. There are a number of vendors of this hardware base. With intense technological and market competition the price and weight of these systems is coming down while the speed and power are rapidly increasing. There is obviously a greatly increased capability for future flexibility with this platform at about the cost of current proprietary data collection systems.

The recent advances in environmentally rugged hand-held MS-DOS computers makes these devices the logical choice for data collection hardware. It can be assumed that due to the plethora of applications that exist for these units that prices will continue to drop while their computing powers expand.

The significant advantage of the MS-DOS hand-held computers as a data collector is the ability to program the unit in standard high level computer language such as C. In theory, any software system developed for use with MS-DOS will operate successfully on these hand-held computers in the field environment. In the future this will become more and more of an advantage. The screen size and storage capacity of current units are the only factors that limit any PC type application from functionally operating on one of these units.

Lack of programming flexibility (a la 41C) is offset by free form data collection and customisable features of the software design.

CEFB is being developed generically for all standard MS-DOS hand-held computers based on what appears to be a fairly consistent storage capacity, screen size, and keyboard.

Total Station Interface:

One of the critical requirements identified for Cadastral Survey data collection is the ability to interface with a multitude of existing survey instruments that may be in use. The CEFB achieves this capability through an incredible amount of cooperation with the Florida Department of Transportation (FDOT, 1990), who has been responsible for a highway-based electronic field book system. Due to this cooperation the technology has been developed that allows CEFB to interface to all commercially available total stations. CEFB will still also allow hand entry of measurements at any time.

The combination of generic PC based data collection software combined with communication to all total stations eliminates any realistic hardware restrictions regarding use of the MS-DOS hardware option for CEFB.

Design Philosophy in Data Collection:

CEFB will have the following basic generic components with regards to data collection:

(1) CEFB will store a file of raw field measurements. In a total station scenario this means storage of horizontal circle, vertical circle, and slope distance in a three-dimensional sense. The ability to measure any combination of the three total station readings will always exist, though default settings will reflect an efficient traverse operation (no distance or vertical circle is recorded when pointing on the backsight, three-dimensional information is recorded when measuring to anything other than the backsight). The default settings will be easily overridden. Data integrity is assured thru protected binary file formatting.

(2) CEFB will not restrict the user in the way a survey is carried out. No predefined number of repetitions will be required. Point naming will be very generic (up to 16 alphanumeric characters), though default naming conventions can be defined for efficiency purposes. A user will be able to move to places unconnected to existing data, though this will create a unique coordinate datum definition until the two survey networks are connected.

Design Philosophy in Data Analysis:

As repetitions are being turned at a setup the user will have the ability to obtain mean and standard deviation of these redundant measurements. The user can reject data, though this data will not be eliminated from the data file but instead flagged as disregarded (analogous to drawing a line through information in a field book).

If the user has defined starting coordinates and direction, the user will be able to recall geodetic (or state plane coordinates) of any station. When existing traverse stations are closed upon the user will easily display conventional latitude, departure, and angular closures.

If record boundary lines are defined, the user will be able to recall distance from a starting point (distance down line), the distance to a defined line (offset to a line), and the distance and direction to any desired point. Upon input of an angle or bearing the user will also be provided with a distance to the defined line.

Astronomy:

Astronomic measurements will be processed real-time, stored in the raw data file, and compared to an existing direction for a line. The user will be able to continue traversing with the result of the astronomic measurements, the existing direction, or some weighted average of the two values.

In short, this design can be described as "Know As You Go", which is essential to a dependent resurvey.

Feature Codes:

One of the critical design factors for BLM Cadastral Survey application in a retracement survey data collector is the ability to efficiently collect descriptive information. This will be accommodated at several levels. First, the user will be able to enter descriptive information free form at any time. In addition for point descriptive data a user-definable look-up table of feature codes will be available in CEFB.

The feature coding will be more than a simple, singular definition if one desires. For example, a feature code for a bearing tree may prompt the user for species, diameter, markings, bearing, and distance. Any or all of the secondary definitions may receive a null entry. The list of feature codes will be readily expandable or customizable by the user.

Initially some thought was given to the potential for full macro assisted word processing capability to exist with the CEFB environment to generate near final *field note ready* descriptions. Further thought about that possibility led to the belief that field operations needed to be streamlined as much as possible in order to be used, and it was unlikely for field people to feel comfortable taking the time to fully annotate descriptions. Offsetting this is the importance of allowing a proper and unbiased description of the actual monument and evidence information. That is, software should not to force the user to fit the description into a predefined form if it is not appropriate. The design of a streamlined yet flexible macro-feature code system will be one of the most difficult parts of the system to implement.

The feature code information will be processed into a data base when the survey measurements are being processed. This information will then be used to help automate production of the official field notes.

Field Notes:

At any time during a survey a user will have the ability to enter descriptive information through the keyboard, just as one would write into a field book. This information will also be extracted and placed into data base form at processing time.

Record to the Field:

The user will also have the ability to store record field notes for the survey in CEFB. This will have been entered through a word processor in the office and transferred to the hand-held PC. This digital information will aide the surveyor in his search for record corners, in computing search positions for missing corners, etc. This information may be encoded using the feature code macro language.

CMM field functionality:

Since a PC is now in the field, the next question raised would be why the entire CMM software system cannot be carried into the field? At the present time storage is the primary problem since computer hard disks, at their present level of development, cannot survive the rigors of field surveying. Also, the physical size of these handheld units limits the size of the display screen. While this is not a problem technically, only a portion of the standard 25 x 80 character screen can be viewed at any given time.

One must then consider which components of CMM are not essential to field surveying operations. It was the mutual consensus of BLM and UM that the survey network analysis routines, centered around the least squares analysis, are not really required. The next items of peripheral importance are the more automated functions for reproportioning entire data sets based on available located corners.

This leaves the following computational items critical in the field (Rodine, et al., 1991):

(1) Proportioning (single, double, broken boundary, etc.) are needed when "unexpected" evidence is found. An example of this is when one is prepared to perform a series of corner moves (positioning lost corners) and evidence is found of a corner during one of the moves. This evidence will possibly affect corner moves which depend on that corner for control.

(2) Geodetic Cogo (see CSTUF program description in Rodine, et al., 1991) is of critical importance in the field for inversing, intersections, midpoints, etc. The screen setup for CSTUF has been modified to fit the hand-held PC and is now fully functional on the field units.

(3) Layout, in the form of corner moves and true line offsets, is a very important field calculation. This now exists in CMM in a program called CMOVE.

Since all of these functions now exist in CMM, are operating geodetically correct, and follow procedures described by the Manual of Survey Instructions (1973), it is simply a matter of altering the user interface for the field unit to make them functional.

Processing - A daily link between CMM and CEFB:

A processor of raw field measurements into abstracted distances, angles, azimuths, and elevation differences has been developed by co-author Hintz for the Florida D.O.T. EFB (FDOT, 1990). When tied to a control file, the software can automatically identify the redundant component of the survey network and save all sideshot observations for later computation.

The redundant data, based on any network geometry, is used to generate initial coordinates for all redundant stations, and a least squares adjustment of the redundant data set is then performed. The sideshot information is then computed from the coordinates generated by the least squares solution.

This entire process is a batch operation, and in CEFB this will mean adding a day's work to the existing data and reprocessing. The user will then employ all of the Cadastral computation tools in CMM to carry out duties in a dependent resurvey.

Prior to the next day's field work, the new coordinates will be re-loaded to the field unit. This will enable one to take the best estimates of station coordinates to the field in an updated fashion.

FUTURE POSSIBILITIES

While it is expected that CEFB development will be evolutionary and ongoing for a number of years, there are several technologies on the horizon which are not part of the current design, but may be considered at a future date.

* Foremost is the technology of a hand writing PC interface instead of keyboard entry. This technology will be extremely usual in a data collection system. It could even allow sketches to return to the electronic field book environment. Until systems are available to test, however, it is difficult to assess whether an interface technology will be a help or hindrance in a particular application.

* Infra red link with the total station. One of the most failure prone and time consuming maintainance aspects of data collection is the cable interface to the survey instrument. It would be desirable in the future to replace this with an infrared digital link (with full data handshake checksum). This may depend greatly on total station manufacturer's development of extremely

compact self contained serial port links that can be installed on any instrument.

SUMMARY

The Cadastral Electronic Field Book (CEFB) has been discussed. CEFB is designed specifically to suit the needs of a surveyor performing a dependent retracement survey in the U.S. Public Land Survey System. The system will allow a user to always know his geodetic position, his relation to other points, his relation to record boundary lines, and perform geodetically correct PLSS computations efficiently in the field. The success of any data collection system is highly dependent upon user acceptance and functionality. Towards this end the process of development of CEFB will, and must, include an evolutionary process that includes user testing, feedback, responsive adaptation, and iterations of these processes.

ACKNOWLEDGEMENTS

Funding for this project was made possible by Cadastral Survey in the U.S.D.I., Bureau of Land Management. The authors wish to thank Bernard W. Hostrop, Chief Division of Cadastral Survey Washington Office, all the BLM Cadastral surveyors who have provided feedback on data collection testing, and an unlistable large number of people at the Florida Department of Transportation who have offered help in this project.

REFERENCES

Blanchard, B.M. (1990), *Utility Program Development: A Digitally Integrated Measurement Management System for the U.S. Public Land Survey System*, M.S. thesis, University of Maine, Orono, ME, 86 p.

Florida Department of Transportation (1990), *Electronic Field Book System - Field Surveying System*, Tallahassee, Fl. 127 p.

Florida Department of Transportation (1990), *Electronic Field Book System - Survey Data Processing System*, Tallahassee, Fl. 76 p.

Rodine, C.J., Wahl, J.L., Parker, B., Blanchard, B.M., and R.J. Hintz (1991), *Progress Report on the Development of an Integrated PLSS Cadastral Measurement Management and Retracement Survey Software System*, Proceedings of the ACSM-ASPRS Annual Convention, accepted for publication.

U.S. Dept. of Interior - Bureau of Land Management (1973), *Manual of Instructions for the Survey of the Public Lands of the United States 1973*, U.S. Government Printing Office, 333 p.

Wahl, J.L. (1990), *Cadastral Measurement Management User's Manual (Beta Release)*, unpublished, 120 p.

CONVERSION OF NAD27 STATE PLANE COORDINATES TO NAD83
STATE PLANE COORDINATES BY A 2-D PROJECTIVE
TRANSFORMATION

Indrajith Wijayratne
School of Technology
Michigan Technological University
Houghton, Michigan 49931

BIOGRAPHICAL SKETCH

Author is an instructor in surveying at Michigan technological University, and teaches upper level surveying courses in the B.S. degree curriculum in surveying. He earned a B.S. degree in physical sciences from University of Sri Lanka, a post graduate diploma in surveying from University College, London, and M.S. degree in geodetic science from the Ohio State University. He is a licensed surveyor in Michigan, and is a member of A.C.S.M. and A.S.P.R.S.

ABSTRACT

With the establishment of the North American Datum of 1983 surveyors are faced with the task of having to convert the state plane coordinates based on the North American Datum of 1927 to the 1983 system. Several transformation models have been suggested for direct conversion without computing the geodetic coordinates on the new datum. This paper looks at yet another direct conversion model which seems more appropriate for the geometry involved. This transformation accounts for the tilt, if any, between old and new projection planes. Test data has shown that this method can yeild accuracies of 0.5 m. or better, in northings and eastings, regardless of the area involved.

INTRODUCTION

The redefinition of the North American Datum has resulted in the use of a world-wide best fitting reference ellipsoid along with the readjustment of horizontal control network. This has substantially reduced the distortions that existed in the previous datum, namely the North American Datum of 1927(NAD27). One of the difficulties encountered by the surveying community, in adopting the North American Datum of 1983(NAD83), is recomputing the state plane coordinates of old control points for which new geodetic coordinates are not known.

There are several methods available for transforming state plane coordinates from NAD27 to NAD83. Following three methods have been established by the National Geodetic Survey(NGS) in their policy letter dated April 29,1983.

1. Rigorous adjustment of original field data to conform with NAD83 and then compute the state plane coordinates on the new system.
2. If the geodetic coordinates on the old datum are known, or computed from the NAD27 state plane coordinates, transform them to NAD83 geodetic coordinates by a rigorous method(eg.LEFTI program developed by NGS), and then compute NAD83 state plane coordinates.
3. Convert NAD27 geodetic coordinates of any new point to NAD83 geodetic coordinates by applying the average differences in latitudes and longitudes, computed from the common points available in the area. NAD83 state plane coordinates can be computed from these geodetic coordinates.

It is obvious that the method 1 yeilds the highest accuracy provided the original field data has some ties to the present system. It is very unlikely that the original field data is available or even compatible with the present system as far as the required precision is concerned. Furthermore, re-organizing this data into the required format to be input into the software will be a very time consuming task.The method 2 is viable but needs the software for rigorous datum transformation and computation of the state plane coordinates on NAD83.The method 3 may not provide the required accuarcy because the average shifts may not be a true representation of shifts at each point. In any case,the success of these methods depends on the existence of common control points in the area.

A direct and convenient method is to use a set of transformation equations which convert the NAD27 state plane coordinates to NAD83 state plane coordinates directly without having to know or compute geodetic positions.The kind of algorithms and software are much simpler and can be run in a low end PC which most surveying companies can afford. The accuracies obtainable by these methods are acceptable for most surveys, and the saving in cost can certainly be a motivating factor.

There have been some attempts to develop some models with varying degrees of accuracy. These transformations are from one plane coordinate system to another which is assumed to be on the same(or parallel) plane. This paper considers a direct conversion model which transforms the coordinates by projecting the points from one plane to another. The two planes need not be parallel.

STATE PLANE COORDINATE SYSTEMS

The two major projections,namely transverse Mercator and Lambert's conformal conic,used in state plane coordinates systems were unchanged, for most part, in implimenting NAD83 state plane coordinates. The standard parallel and central meridian values were unchanged, except in a very few instances(eg. Michigan's south and central zones).

Units of the coordinates on NAD83 are in meters even though some states will continue to use feet.

The plane for the coordinate system is developed by unfolding a cone which intersects the datum ellipsoid along two standard parallels (Lambert's conformal conic), or a cylinder which intersects the ellipsoid along two small circles parallel to the central meridian(transverse Mercator). Exception to this is the oblique Mercator used in Alaska. Any change in the ellipsoidal parameters or the defining constants of the projection will generate a different plane. This is the case in changing from NAD27 state plane coordinates to NAD83 state plane coordinates. These two planes may have a slight tilt depending on the parameters that were changed. For example, in Lambert's conformal conic, the cone has a different angle at the vertex when the standard parallels are changed. Even if the defining constants such as standard parallels were not changed numerically, their topographic positions have changed due to the change of datum.

A DIRECT TRANSFORMATION

A direct transformation of state plane coordinates involves the following:

1. A rotations necessary to correct for any difference in orientation between the two systems
2. Translations to correct for any difference in origin, sometimes due to different false values imposed
3. A scale correction due to different size datum ellipsoids, change in scale factors due to change in projection parameters, and differences in units used (eg. feet and meters).

Even though each projection in itself is conformal (exception to this is that the transverse Mercator used in NAD27 for the conterminous U.S.was slightly non conformal) conformality may not be preserved in a transformation from one coordinate system to another due to distortions in the the first coordinate system, any tilt between the planes,and the deficiencies in the model. When the model parameters are determined in a least squares solution the magnitude of the residuals reflect the distortions in the control points used and the deficiency in the model. The residuals can be made smaller by increasing the degree of the polynomials used(Vincenty, 1987), but this may hide the distortions present in the control points. The distribution of common control points is another factor that may affect the residuals.

The following transformation models have been suggested or considered by several individuls. The notations used in all the equations are as follows:

X_{27}, Y_{27} = Easting and northing, respectively, of NAD27 state plane coordinates
X_{83}, Y_{83} = Easting and northing, respectively, of NAD83 state plane coordinates
a,b,c,d etc.= coefficients of transformation equations

Two dimensional Conformal(Helmert or Similarity) Transformation(Vincenty):

$$X_{83} = X_o + aX_{27} + bY_{27}$$
$$Y_{83} = Y_o + aY_{27} - bX_{27}$$

Two dimensional Affine Transformation(Shrestha and Dewitt, 1989):

$$X_{83} = a_o + a_1 X_{27} + a_2 Y_{27}$$
$$Y_{83} = b_o + b_1 X_{27} + b_2 Y_{27}$$

Second degree Conformal Polynomial transformation(Moffitt and Jordan):

$$X_{83} = a_o + a_1 X_{27} + a_2 Y_{27} + a_3 (X^2_{27} - Y^2_{27}) + a_4 (2X_{27}Y_{27})$$
$$Y_{83} = b_o - a_2 X_{27} + a_1 Y_{27} - a_4 (X^2_{27} - Y^2_{27}) + a_3 (2X_{27}Y_{27})$$

Vincenty(Vincenty, 1987) also suggested the same model but with two additional third degree terms. Shrestha(Shrestha, 1987) showed that decimeter level accuracy may be obtained for a limited area by using the equations obtained by fitting straight lines to the differences of eastings and northings obtained through DMA Multiple rgression Model. Shrestha and Dewitt(Shrestha and Dewitt, 1989) tested the affine transformation model and concluded that an average absolute precision of 0.6 meters may be obtained even for a large area like a full zone.

The model considered in this paper is a two dimensional projective transformation model. This transformation is quite common in photogrammetric applications where photo coordinates have to be transformed into map(or ground) coordinates in the process of rectification. The model, as given below, projects points on one plane to a nonparallel plane while at the same time accounting for any distortions in the first set of coordinates such as tilt distortions present in an aerial photo. This may be appropriate for conversion of state plane coordinates as the two projection planes on two datums may have a slight tilt as shown earlier. Strictly, the projection of points must be done from a single point(perpective),but the equations are general in nature, and simply express a projection from one plane to another non parallel plane (Wolf,1983). This transformation is not conformal, and therefore, two scale changes are allowed.

The transformation has eight parameters as shown below (Wolf, 1983).

$$X_{83} = \frac{a_1 X_{27} + b_1 Y_{27} + c_1}{a_o X_{27} + b_o Y_{27} + 1}$$

$$Y_{83} = \frac{a_2 X_{27} + b_2 Y_{27} + c_2}{a_o X_{27} + b_o Y_{27} + 1}$$

This model, unlike the previous models, is non-linear in the transformation coefficients, and therefore should include iterations in a least squares solution for evaluating them. One could also use a linear approach by rearranging the equations as shown below (Manual of Photogrammetry).

$$a_o X_{27} X_{83} + b_o Y_{27} X_{83} + X_{83} - a_1 X_{27} - b_1 Y_{27} - c_1 = 0$$
$$a_o X_{27} Y_{83} + b_o Y_{27} Y_{83} + Y_{83} - a_2 X_{27} - b_2 Y_{27} - c_2 = 0$$

This will be a mixed model, for the purpose of least squares adjustment, of the form

$$F(L_a, X_a) = 0$$

where L_a = adjusted observations (state plane coordinates) and
X_a = adjusted parameters (coefficients of equations)

Both these approaches resulted in identical coefficients and residuals. Non linear model converged in less than 5 iterations, in most cases, when proper approximate values were chosen. Both of the following initial approximations were satisfactory.

$$a_1 = a_2 = 1, \quad b_1 = b_2 = a_o = b_o = 0, \text{ and}$$
$$c_1 = \text{Ave}(X_{83}) - \text{Ave}(X_{27})$$
$$c_2 = \text{Ave}(Y_{83}) - \text{Ave}(Y_{27})$$

or
$$a_1 = a_2 = b_1 = b_2 = 1, \quad a_o = b_o = 0, \text{ and}$$
$$c_1 = \text{Ave}(X_{83}) - \text{Ave}(X_{27} + Y_{27})$$
$$c_2 = \text{Ave}(Y_{83}) - \text{Ave}(X_{27} + Y_{27})$$

A minimum of four common points are needed for the computation of transformation coefficients, but more points should always be used.

TEST RESULTS

Two areas in the north zone of Michigan were selected as test sites. The following results obtained for three models, namely 2-demensional affine transformation, second degree polynomial transformation (Moffitt and Jordan), and 2-D projective transformation are shown for the purpose comparison. The residuals(X/Y) were obtained from the least squares solution used for evaluating the transformation parameters. The descrepancies(X/Y) are the differences between transformed coordinates and the NGS published values. All values are in meters.

Area-1

        Approximate dimensions = 45 mi X 30 mi
        No. of common points used = 13
        No. of points transformed = 121

Model	Max abs residual	Ave. abs residual	Max. abs descrep.	Ave. abs descrep.
Affine	0.42/0.32	0.11/0.11	0.96/1.01	0.23/0.22
2nd deg. Poly.	0.36/0.34	0.21/0.12	3.14/1.09	0.26/0.26
2D-Project.	0.42/0.34	0.19/0.11	1.13/2.09	0.24/0.23

Area-2

        Approximate dimensions = 90 mi X 120 mi
        No. of common points used = 36
        No. of points transformed = 258

Model	Max abs residual	Ave. abs residual	Max. abs descrep.	Ave. abs descrep.
Affine	1.18/1.01	0.36/0.37	1.27/1.59	0.30/0.36
2nd deg. Poly.	1.16/1.00	0.53/0.47	1.36/1.35	0.47/0.49
2D-Project.	0.52/0.68	0.21/0.20	0.70/0.96	0.23/0.26

CONCLUSIONS

The model considered is just another method for direct transformation of state plane coordinates from NAD27 to NAD83. It has shown encourging results for the study area. Following conclusions were made by analyzing the results.

1. The model considered yeilded comparable residuals and precisions in transformed coordinates in both test areas, indicating that the extent involved has no effect on the transformation.
2. All three models resulted in comparable precisions when extent of the area involved is small.
3. The increase of the number of common control points in a given area beyond 10 did not make any substantial improvements in the results.
4. Further investigation is needed to make a firm conclusion.

Even though the theoretical model assumes a finite pojection center it can be used for projection from one plane to another. The model has been used for transforming comparator coordinates measured on aerial photographs to a coordinate system based on fiducial points(Wolf, 1983).

All the common points used in this study were considered to be of the same quality. But, this is certainly not the case, and a weighting factor may be introduced into the least squares solution in evaluating transformation coefficients. Vincenty(Vincenty,1987) has suggested a smoothing out of residual errors at transformed points in order to distribute the effect of distortions.

ACKNOWLEDGEMENTS

Author sincerely acknowledges the support given by Mr. David Zilkoski of the National Geodetic Survey in writing this paper by providing the necessary data and software for state plane coordinate computations.

REFERENCES:

1. American Society of Photogrammetry and Remote Sensing (1980), "Manual of Photogrammtery", 4th edition
2. Moffitt, Francis H. and Jordan, Susan L. (1987), "Conversion of State Plane Coordinates from NAD27 to NAD83 by Second-Degree Conformal Transformation", ACSM Bulletine, No.106,pp. 13-20
3. Shrestha, Ramesh L.(1987),"Coordinates Transformation from NAD27 to NAD83", Surveying & Mapping, Vol. 47, No. 4,pp. 295-300.
4. Shrestha, Ramesh L. and Dewitt, Bon A.(1989),"An Evaluation of NAD-27 to NAD-83 Plane Coordinate Transformations for the State of Florida", Surveying and Mapping, Vol. 49, No. 4,pp.179-183

5. Stem, James. E.(1989),"State Plane Coordinate Systems of 1983", NOAA Manual NOS NGS 5
6. Vincenty, T.(1987), " Conformal Transformations Between Dissimilar Coordinate Systems", Surveying & Mapping, Vol.47, No. 4, pp.271-274
7. Vincenty, T., Personal Communications
8. Wolf, Paul R.(1983), "Elements of Photogrammetry", 2nd edition, McGraw-Hill Book Company
9. Uotila, Urho A., Class Notes, Ohio State University

GEODETIC CONTROL:
TRADITIONAL TECHNOLOGY TRANSFORMED FOR TODAY

Cari Winslett
310 Lakeshore Drive #20
Lake Park, Florida 33403

Robert J. Pearsall
1742 Shoreside Circle
West Palm Beach, Florida 33414

ABSTRACT

The perspective of Geodetic Surveying rises beyond the traditional monument restoration, section retracement, and other field related applications. The use of precision input mapping techniques, enhanced by sophisticated measurement capturing,(GPS, total stations, lasers, EDM, etc.), allows cartographers, cadastralists, and other map creators unlimited reliable information essential for GIS mapping. The availability of application specific information provides the power to analyze, manipulate, and produce highly accurate maps suitable for multi-purpose cadastre and GIS use by multi-disciplined personnel far removed from actual field surveyors.

Using a coordinate based skeletal framework provides the necessary control for the unity of the physical world and the "legal description." The product is a composite of graphic information and non-graphic data capturing. A coordinate based GIS map allows flexibility for referencing chronologically dissimilar information. The rigidity of certifying monuments removes the liability aspects associated with accessing electronic data. Using a coordinate system removes the arbitrary individualistic decision of geographic placement in the creation of digital maps. Reliability and perpetual maintainability is paramount to a successful GIS.

Automated mapping systems in the governmental sector include property assessment. In the past many jurisdictions only used maps for locational purposes, today there is a need for spatially and descriptively accurate maps. Achieving this tasks necessitates the use of a controlled, dependable monumentation system. The creation and maintenance of cadastral maps utilize an amalgamation of cross-time data.

The actualization of an accurate map base in the electronic medium facilitates the skills, knowledge, and product of the geodetic surveyor.

Surveyors must have an active role in the creation of a successful Geographic Information System. As GIS becomes more of a science than an art form, the "product" used for the creation of the skeletal framework should reflect the intense precision used by the technical experts(surveyors, photogrammetrists and remote sensing technicians.) The field surveyor may not always realize that his work becomes the basis for geometrically and topologically accurate GIS.

For years the industry provided sophisticated methods of data capturing, the surveyors would utilize this data, but the GIS community did not. The union of the precision non-graphical data and precision geometric data will produce the ultimate GIS.

The removal of the barriers for usage of precision geometric data lie in the flexibility of computer equipment and technology. In the past, coordinate geometry was not easily accessed by individuals without specialized training. The cumbersome, complex, and difficult training acted as a deterrent for precision input(COGO) usage. The personnel computer systems and programs of today allow direct downloading of field data to an environment conducive for utilization of this information.

Data classified as chronologically incompatible was once deemed impossible to assimilate in an accurate mapping base, now, with the careful blending of multi-disciplined skills this data becomes both reliable and compatible. A rigid network of perpetually maintained control assures correct geographical placement of geometric structures. Older information obtained from antiquated techniques provide invaluable reference information. The new, sophisticated methods of data capture must then return to the senior reference information for proper physical placement. The ideology that the "real world" and the "legal description world" do not match, is not true. With a blend of technical staff, experience, and skills, the two worlds do converge.

The keys to a successful application of a geodetic network are:
- CHOOSE THE BEST AVAILABLE CONTROL NETWORK
- PERPETUALLY MAINTAIN AND UPGRADE MONUMENT CONTROL
- REFERENCE OLD SURVEYS TO NEW COORDINATE INFORMATION
- UTILIZE OLD SURVEY FIELD NOTES
- SET STANDARDS AND CONTINUE THEIR USAGE
- MAINTAIN CONSISTENCY

Surveyors face liability issues. The digital or electronic medium will not cause liability problems if properly used. With the application of a coordinate based system, individual and personal liability issues will diminish. A unified, common base for all surveyors to reference will reduce the arbitrary, individualistic determinations that lead to liability conflicts. Once referenced to a standardized grid, a point of beginning on an older survey becomes intelligent. Hiatuses or overlaps may not occur at all if carefully referenced to the chosen, standard coordinate system. Remember, bearing bases in most surveys are relative. With the technology available today, this relativity will disappear.

CONCLUSION

GIS needs coordinate based mapping. Your skills, knowledge, and applications of new techniques is a requirement. Without the support of the geodetically interested factions GIS mapping cannot succeed.

THE 1989 MT. McKINLEY, ALASKA, GPS EXPEDITION

Jeffrey F. Yates
AeroMap U.S., Inc.
2014 Merrill Field Drive
Anchorage, Alaska 99501

ABSTRACT

Early in 1988, planning was begun by a group of individuals, private companies, the University of Alaska, and government agencies with the goal of safely and accurately remeasuring the summit elevation of Mt. McKinley, Alaska. The tallest mountain on the North American Continent was surveyed using state-of-the-art Global Positioning System (GPS) surveying equipment and techniques. The significance of this project lies in the combination of using advanced surveying technology on a world-class climbing expedition, in a remote and harsh environment.

INTRODUCTION

The project was divided into three main phases.

Phase I consisted of the collection of gravity data which utilized LaCoste-Romberg gravity meters on and around the mountain via U.S. Army Chinook helicopters and ground vehicles.

Phase II had an eight-man climbing expedition placing two GPS receivers on the summit on two separate days. LaCoste-Romberg gravity measurements were positioned using the GPS receivers. Additional gravity measurements were completed using a Worden gravity meter.

Phase III performed additional LaCoste-Romberg gravity surveys around the base of the mountain within the boundaries of Denali National Park utilizing a Bell Jet Ranger helicopter.

Phase I was completed in May, 1989, when the U.S. Army granted permission to fly aboard their Chinook high altitude helicopters on two separate days. Gravity measurement loops were made between their staging area and the mountain.

In June 1989, Phase II was accomplished when seven of the eight climbers successfully carried two GPS receivers to the summit of Mt. McKinley, recording data on two separate days.

Simultaneous measurements were made by a support team equipped with GPS receivers placed at selected control stations around the base of Mt. McKinley.

Phase III was completed in June of 1990, when a LaCoste-Romberg gravimeter was used to survey gravity loops around the base of the mountain. This gravity data will be used by the National Geodetic Survey to better define the 'geoidal' height needed to finalize the summit elevation derived from the GPS observations.

PROJECT BACKGROUND

The currently published summit elevation of Mt. McKinley of 20,320 feet (6,194 m) was measured in 1953 by Bradford Washburn, in conjunction with Commander Howard Cole of the U.S. Coast & Geodetic Survey. Washburn utilized optical surveying instruments and trigonometric leveling - the most accurate method for such remote measurements at that time. The mountain was remeasured in 1977, again by Washburn. (Note: Dr. Washburn is credited with being the first to pioneer the West Buttress route in 1951, which has since become the premier route up the mountain.)(Waterman, 1988).

In 1986, national and international attention was focused in the Himalaya mountains when controversy arose over whether K2, long established as the world's second highest mountain, might actually be taller than Mt. Everest. A preliminary Doppler satellite measurement taken by University of Washington professor and astronomer George Wallerstein suggested that K2's summit elevation **might** be 29,064 feet (8,859 m), or even 29,228 feet (8,909 m), either of which would have been higher than the 29,028 foot (8,848 m) Mt. Everest. Although later satellite measurements proved Everest to be the taller of the 2, the controversy helped initiate the way for future use of satellite technology to determine more accurate heights of major peaks around the world. (Krakauer, 1990).

In 1988, the Land Surveyors Association of Washington performed a well planned and executed GPS survey of Mt. Rainier, whereby they determined the new summit elevation to be 1.1 feet higher than the old calculation of 14,410 feet (4,392 m). This same year, planning was underway to use this same technology on Mt. McKinley, Alaska.

Because of its great height and remote location, it was determined early in the planning stages that the best way to safely measure the summit elevation of Mt. McKinley would be to organize a climbing expedition.

Spending three to four weeks climbing the mountain would allow the team to 'acclimatize' to the rarified air above 20,000 feet (6,096 m). The use of a helicopter was ruled out because 20,300+ feet (6,188 m) Mean Sea Level is beyond the serviceable ceiling for the majority of rotary-wing aircraft.

TECHNICAL PREPARATIONS

The centerpieces of the expedition were the six L-XII receivers supplied by Ashtech Inc. The GPS receivers were carrier-phase, single-frequency, self-contained units. Each receiver, along with its microstrip antenna, weighed 15 pounds. The only modifications made to these off-the-shelf models, were the removal of the internal batteries on the two receivers designated to go to the summit and the removal of the ground-plane accessories and tripod attachment brackets from the standard Ashtech antennas. To power the receivers, lithium batteries were chosen for their light weight and ability to operate in temperatures down to -40° Fahrenheit. Special arctic cables were fabricated by Geonex/Itech, Inc. to ensure proper cold weather operation interface between the batteries, antennas, and receivers.

Four gravity meters were utilized during the three Phases. Two LaCoste-Romberg gravimeters were provided by the National Geodetic Survey. These highly sensitive, temperature controlled instruments were used to establish "gravity stations" at Talkeetna, plus the 7,100 foot (2,164 m) , 14,200 foot (4,328 m) and 17,200 foot (5,243 m) elevations on Mt. McKinley. One additional LaCoste-Romberg meter was supplied by Geonex-Itech for Phase III.

Due to the bulk, high weight, and power requirements of the LaCoste-Romberg gravimeters, a lighter expedition weight gravity meter was chosen for use by the climbing team. A ten-pound Worden gravimeter was leased from Neese Instrument Company. This gravimeter was used to measure a gravity loop from the base camp at 7,100 feet (2,164 m) to the summit and back. Unfortunately, its disadvantages included lower accuracies and the need to be reset quite often, as its scale only allowed for a 2,000 foot (609 m) change in elevation. The LaCoste-Romberg gravity stations created in May were used as check-points for the Worden drift error during the climb.

Communication between the climbing team and the "ground" support team was accomplished via two UHF radios supplied by Denali National Park using National Park Service frequencies. Citizen Band radios were used for on-mountain communications between team members and other climbing groups. A VHS-S video camera was brought along to record events of the expedition.

Being that the true summit (highest piece of rock) of the mountain is encased in a mantle of centuries-old ice of unknown depth, it was decided to place a USGS monument at the highest point of the ice-summit. Berntsen, Inc. manufactured an extremely durable four-inch diameter bronze/magnesium alloy cap, attached to a one-half inch stainless steel rod via a special locking mechanism. An ice-auger type bit was attached to the first rod for drilling the monument permanently into the icecap.

Specifications and recommendations for the field planning of the GPS and gravity measurements were sought from the National Geodetic Survey in Rockville, Maryland. John Oswald of Geonex-Itech was the on-site expert involved in the coordination and planning for final placement of the GPS receivers. Four of the GPS receivers were placed on one Second-order and three First-order stations near the base for simultaneous measurement by the support team with the two summit receivers. The benchmarks occupied by these GPS receivers were:

             Benchmark J-126, NGS First-order
             Wonder Lake IGEHU, NGS First-order
             GPS-9, ADOT/PF, GPS Second-order Station
             Benchmark S-126, NGS First-order

Communication was made with Dr. Bradford Washburn of the Boston Museum of Science regarding which benchmarks to use, especially for the purpose of comparing values with those of his obtained in 1953. Permits were required for the various types of work to be performed in Denali National Park and were acquired as necessary.

CLIMBING PREPARATIONS

Mt. McKinley is one the coldest mountains on earth. Additionally, due to the lower barometric pressures of the higher latitudes, McKinley is considered to be physiologically 2,000 to 3,000 feet (609 m to 914 m) **taller** than it actually is, as compared to equivalent elevations in the Andes or Himalayas. Its vertical rise from the surrounding terrain of over 18,000 feet (5,486 m) is greater than that of Mt. Everest, which rises from 14,000 feet (5,486 m) to just over 29,000 feet (8,839 m).

Inherent natural conditions on Mt. McKinley which can affect any climbing expedition consist of ambient air temperatures that, even in June, can reach down to -40° Fahrenheit. Windstorms may have velocities in excess of 100 miles per hour which can last a week or longer, bringing the wind-chill factor down to or below -148° Fahrenheit. (Davidson, 1986). Exposed skin will freeze in literally seconds. Blizzards can drop up to three feet of new snow per day and last for a week. Following newly fallen snows are the dangers associated with avalanches. Icefalls from shifting of the glaciers can occur at any time without warning. Falling into a snow covered crevasse, some of which are large enough to swallow a bus, is an ever present danger. On the wind-swept upper mountain, the trail can be reduced to shear blue-ice. Physiological problems include hypothermia, dehydration, physical and mental fatigue, lassitude, and carbon monoxide from cooking stoves. At the higher altitudes, sicknesses such as cerebral or pulmonary edemas (fluid build-up) can kill a climber in as little as two hours if they cannot be brought down to lower elevations. With approximately 60 deaths credited to Mt. McKinley, climbers soon learn to give the mountain the respect it demands.

In addition to the technical details of the project, preparations for the climbing expedition were taking place simultaneously. Being that it was planned to take GPS measurements on the summit on two separate days, the climbing team needed to be large enough to split into two groups. Additionally, statistics show that on an average only 50% of those attempting the climb actually achieve the summit. It was decided an eight-man team would be best for the purpose of the climb. The climbing team consisted of:

             Brian Clark   Physical Trainer/Nutritionist
             Ronald Cothren   Project Coordinator

Michael Dagon	Expedition Leader
Drow Millar	Videogragher
Stephen Parker	Medical Advisor/EMT
Peter Richter	Communications
Vernon Tejas	Assistant Expedition Leader
Jeffrey Yates	Assistant Project Coordinator

Dagon, Parker, and Yates previously climbed Mt. McKinley in 1987. Tejas was making his 22nd climb of the mountain. In 1988, Tejas became the first to climb the mountain solo in winter and survive.

Preparations included physical training that began at least six months before the actual climb. Technical rope work, safety and emergency procedures were repeatedly practiced before arriving on the mountain and continued during the climb. Food for eight men for 28 days was purchased and packaged. Tents, sleeping bags, ropes, climbing hardware, stoves, fuel, snowshoes, crampons, shovels, clothes and many miscellaneous but necessary items had to be purchased, sorted, and packed. Medical kits were prepared and rechecked. Various strategies were discussed on how to climb the different terrain encountered on the mountain.

SEQUENCE OF EVENTS

PHASE I

For most of the past decade, the Denali Mountain Research Center (DMRC), headed by Dr. Peter Hackett in association with the University of Alaska, has set up a medical camp at the 14,200 foot (4,328 m) level on the mountain. The purpose of this facility is to study the effects of cold and high-altitude on human physiology. The U.S. Army transports in the several tons of medical gear and supplies using specially equipped Chinook helicopters. It was aboard these helicopters that Yates and Cothren were able to perform the LaCoste-Romberg gravity loops. Both loops began and ended at a First-order benchmark located at the Talkeetna Flight Service Station (FSS), located approximately 60 air miles from the mountain. On May 1, 1989 Yates completed a gravity loop from the benchmark at the FSS, to the DMRC Medical Camp at 14,200 feet (4,328 m) and back to the FSS. The following day, the weather permitted both Yates and Cothren to complete a gravity loop from the FSS to 14,200 feet (4,328 m) to 17,200 feet (5,243 m) and back to the FSS. Bottled oxygen was used to alleviate altitude sickness due to the abrupt ascent to altitude. (The expedition members did not carry oxygen during the June climb. "Expedition" style climbing generally allows more time for the body to acclimate to the lower oxygen levels at high altitude, decreasing the chances for acute mountain sickness.)

Paul Brooks, USGS, Don D'Onofrio, NGS, and Cothren performed a LaCoste-Romberg gravity loop in May from Talkeetna to a station located at Toklat on the north side of Mt. McKinley, returning to Talkeetna. During this loop additional measurements were made along the Parks Highway outside of Denali National Park on various First- and

Second-Order benchmarks. Brooks returned later in the season to complete the loop from Toklat to Wonder Lake and back to Toklat.

PHASE II

On May 31, all expedition members arrived at the Talkeetna Air Taxi facility at Talkeetna, ready to be flown onto the mountain. However, due to inclement weather, flights did not begin until the morning of June 2nd. Four flights were required to transport all eight members to the base camp at the 7,100 foot (2,164 m) level. Brooks accompanied the first planeload to base camp. Using 1 of the LaCoste-Romberg gravity meters, Brooks recorded the third gravity survey loop between the Talkeetna FSS and Mt. McKinley, thus completing Phase I of the project. Brooks returned to Talkeetna later that day.

The approximate 1,100 pounds of gear were divided up among the eight members and the climb towards the summit, which was 15 miles north and 13,200 feet (4,023 m) up, began. Due to the weight, two to three trips were required between each camp to transport the supplies and scientific gear up the mountain. The team traveled expedition style, staying at the recommended pace of climbing no more than 1,000 vertical feet per day. On June 16th, the team reached the DMRC Medical Camp at 14,200 (4,328 m) feet. Even with staying on the conservative climbing schedule, 1 member, Clark was not able to avoid being afflicted with altitude sickness. He remained at this camp until the team returned from the upper mountain. In addition, during the 1st of three trips from 14,200 feet (4,328 m) to 17,200 feet (5,243 m), Cothren physically overextended himself and was forced to return to the safety of the Medical Camp with the aid of Parker. After being treated for hypothermia, Cothren was able to continue the climb two days later.

On the 16th of June, the team left the 14,200 foot (4,328 m) camp and reached the 16,200 foot (4,938 m) level late in the evening, taking over seven hours to go the 2,500 feet (762 m) horizontal and 2,000 feet (609 m) vertical distance. Early the next morning, the team departed and climbed along the exposed West Buttress ridgeline to the 17,200 foot (5,243 m) camp. This is a knife-edge route that has shear drops of 2,000 feet-3,000 feet (609 m-914 m) on either side.

Normally, most climbers make the high camp at 17,200 foot (5,243 m) their staging area for attempts at the summit. But knowing that the period of time when four or more of the Navstar satellites were visible ten degrees or more above the horizon was between 1:00 p.m. and 5:00 p.m., the expedition members planned to stage the summit attempts from Denali Pass. A minimum of four satellites are needed for three-dimensional determination of coordinates. The short duration of the satellite "window" was of major concern, due to the fact that it was necessary to match the short satellite window with a weather window. Unfortunately for the expedition, the weather strictly dictates when climbers can and cannot move on the mountain. The expedition members planned beforehand that the travel time between high camp and the summit could be shortened by up to eight hours when starting out from the 18,000 foot (5,486 m) Denali Pass. Timing would be critical in that extra time would be needed because of the additional weight of the scientific gear and the need to take a gravity measurement during the ascent. Additionally, the monument would have to be installed and still have enough time within the satellite/weather window to set up the receiver and record two hours of continuous GPS data.

The only problem with camping at Denali Pass is that it is a gamble. At 18,000 feet (5,486 m), Denali Pass is the saddle situated between the North and South Peaks of Mt. McKinley. It has a notorious reputation for being a tremendous natural wind tunnel, created by the venturi effect between the two peaks. Wind speeds have been estimated to exceed 150 miles per hour in this area (Davidson, 1968).

After carrying one load up to the 18,000 foot (5,486 m) camp, the expedition departed for the higher camp June 19th. After an exhaustive climb up one of the most dangerous sections of the route, the team arrived at the chosen camp site at the head of the Harper Glacier in blizzard conditions. The high camp then had to be constructed. Snow blocks were cut to build walls approximately eight feet tall and three feet to four feet thick to protect the tents from the winds of Denali Pass. The windwalls were shaped to resemble the prow of a ship pointed towards the direction of the prevailing winds in order to minimize the erosive effect of the winds.

Two days after making camp at Denali Pass, the weather allowed the first summit team to try for the summit. The GPS base team was notified that Team One was on its way. To ensure simultaneous GPS observations at the summit and at the base stations, it was decided early on that the base stations would observe every day during the time when the summit teams could be on the summit, regardless of positive radio contact.

Team One, consisting of Dagon, Millar, Tejas, and Yates reached the summit at 1:00 p.m. on June 21 in clear -10° Fahrenheit weather, with gusty winds. There was one large lenticular-shaped cloud that hovered just above a large portion of the Alaska Range that everyone kept a very close watch on. The USGS monument was augered into the summit to a depth of four and one-half feet. The GPS receiver, lithium batteries (one primary and one backup), antenna and connecting cables were assembled and secured with ice screws and cords. Shortly after the power was turned on, the receiver locked onto five satellites. A gravity measurement was taken. GPS data was collected for two hours before Team One left the summit. The GPS receiver continued to collect data until the batteries expired. The lenticular cloud did descend on the summit shortly after the team left.

Team Two, consisting of Cothren, Parker, Richter, and

Tejas, made a summit attempt on the morning of June 22. Low visibility and increasing winds forced them back to camp after several hours. The storm would keep the seven climbers confined to the two small tents for two days except when it was necessary to repair the worn windwalls.

By the 24th, the storm subsided and Team Two started up on their second attempt. The visibility was approximately two to three miles, the temperature was -15° Fahrenheit and no wind. Aside from getting cold feet, Team Two reached the summit without any problems and set up the second GPS receiver. After digging the drifted snow out from around the connectors, fresh batteries were connected to the GPS receiver left by Team One. Both receivers began tracking and recording the signals from five satellites soon after they were turned on.

Both receivers were set up for the purpose of redundancy. Each receiver had its own antenna and power supply. The antenna for the second receiver was placed directly on the summit monument whereas the first antenna was placed a short, measured distance from the monument.

During the two hours of GPS data collection, Tejas, who carried along a parasail, decided the conditions were satisfactory enough to fly from the summit. In doing so, he became the first person to parasail off the summit of Mt. McKinley. He flew down to approximately the 19,000 foot (5,791 m) level, stashed his parasail and climbed back up to the summit. By the time he arrived back at the summit, the GPS data recording was completed but the winds had picked up considerably. Visibility throughout the day had been up to two miles but was now reduced to several feet. The team packed the receivers, roped up, and began their descent in the storm. On the exposed summit ridge the only way to keep from stepping off the 8,000 foot (2,438 m) cliff at times, was to grope along using their ice-axes to keep their bearing along the ridgeline. The team eventually made it to Denali Pass and the relative shelter of the windwalls and tents. It was later learned that this was the worst wind storm of the summer season, damaging tents of other climbing teams.

Early morning on June 25th, the team packed up and descended to the 17,200 foot (5,243 m) camp. The supplies and tent that had been left there had to be dug out and packed for the continued descent to 14,200 feet (4,328 m). After being reunited with Clark, the team was ready to continue the descent by 4:00 the following afternoon. Aided by the darkless Alaska summer nights the group pushed on through the 'night' to arrive at base camp 12 hours later at 4:00 a.m. on June 27. Aside from being highly motivated to get home at that point, it was also safer to travel the lower glaciers at night during the 'warm' summer season. During the long days, the snow bridges covering the extensive crevasse fields begin to collapse and disappear, sometimes without advance notice. At night, when the sun dips just below the horizon, the air temperature will drop below the freezing point, solidifying the snow bridges.

Expedition members began flying out with various air taxis services by 7:30 a.m. and all were back in Talkeetna by 8:00 p.m. that evening, ending Phase II of the project.

PHASE III

Phase III of the project was completed on June 11, 1990 when Alaska Helicopters donated two days' use of a Bell Jet Ranger in order to perform two LaCoste-Romberg gravity loops. Geonex-Itech provided the gravity meter and inertial positioning equipment necessary to complete the survey. Steve Smith piloted the aircraft while Terry Mesick operated the inertial equipment and performed the gravity measurements. Due to the remote and harsh environment surrounding Mt. McKinley within Denali National Park, the use of a helicopter was determined to be necessary. The appropriate permits were obtained from the National Park Service.

Two separate loops were completed to provide additional gravity data for use in determining a better geoid model around the base of the mountain. The first gravity loop was completed to the south of Mt. McKinley, starting from Station A-109. This station was established as a gravity station in May, 1989, by Cothren and Yates. The loop continued with an Geonex-Itech GPS Station, also located at Talkeetna, to four BLM Doppler Stations established in 1988, to stations on the Kahiltna and Yenta Glaciers. Coordinates for these glacier stations were obtained by inertial navigation equipment. The loop was then closed back to Station A-109 in Talkeetna.

The second gravity loop was completed on the north side of Mt. McKinley. This loop began at USC & GS First-order Benchmark J-126. Additional measurements were made on Benchmarks Muddy, Birch, Pastel and Swallow, closing back on Benchmark J-126.

CONCLUSION

To date, the **preliminary** data processing suggests that the summit elevation is approximately 20,306 feet (6,189 m), down 14 feet (4.27 m) from the currently published elevation of 20,320 feet (6,194 m). This cannot be verified until all of the gravity data collected has been processed and the shape of the geoidal surface in the Mt. McKinley area has been defined to NGS standards. Additionally, many of the ice-capped peaks in the Alaska Range have been rounded to the nearest ten feet because of the unstable nature of ice. The preliminary results for the three separate GPS receiver observations are:

	Latitude	Longitude	Elevation
(Day 1)	63 04'8.764"	151 00'22.916"	20,307.30 feet
(Day 2)	63 04'8.774"	151 00'22.971"	20,305.90 feet
(Day 2)	63 04'8.764"	151 00 23.016"	20,306.47 feet

The data collected from the three phases is currently being

processed into NGS "Blue Book" format. Final approval for a new elevation of the summit of Mt. McKinley will be announced by the NGS.

ACKNOWLEDGEMENTS

The following organizations participated in the project:

>AeroMap U.S., Inc., Anchorage, Alaska
>Alaska Helicopters, Inc., Anchorage, Alaska
>Alaska Mountaineering and Hiking, Anchorage, Alaska
>Alaska Department of Transportation, Juneau, Alaska
>Ashtech Inc., Sunnyvale, California
>Berntsen, Inc., Madison, Wisconsin
>Bureau of Land Management, Anchorage, Alaska
>Denali Medical Research Center, Anchorage, Alaska
>Genet Expeditions, Anchorage, Alaska
>Geonex-Itech, Anchorage, Alaska
>Museum of Science, Boston, Massachusetts
>National Geodetic Survey, NOAA, Rockville, Maryland
>National Park Service, Denali National Park, Alaska
>NCS International, Anchorage, Alaska
>University of Alaska, Anchorage, Alaska
>U.S. Army, Fort Wainwright, Alaska
>U.S. Geological Survey, Anchorage, Alaska

REFERENCES

Cothren, Ronald G., and Yates, Jeffrey F., (1990) "Taking GPS to New Heights," P.O.B. Vol.15, Number 6, pp.40-50.

Davidson, Art, (1986),"Minus 148 Degrees: The Winter Ascent of Mt. McKinley", Cloudcap Press, Seattle, Washington, pp.214-215.

Krakauer, Jon, (1990), "Eiger Dreams, Ventures Among Men and Mountains", Lyons & Burford, New York, New York, pp.116-129.

Waterman, Jonathan, (1988), "High Alaska, A Historical Guide to Denali, Mt. Foraker and Mt. Hunter", The American Alpine Club, New York, pp.84-85.

A PRIORI ESTIMATES OF STANDARD ERRORS
OF LEVELING DATA

David B. Zilkoski
NAVD 88 Project Manager
Vertical Network Branch
National Geodetic Survey
Rockville, Maryland 20852

BIOGRAPHICAL SKETCH

David B. Zilkoski received a B.S. degree in Forest Engineering from the College of Environmental Science and Forestry, Syracuse, New York, in 1974, and an M.S. degree in Geodetic Science from the Ohio State University in 1979. He has been employed by the National Geodetic Survey (NGS) since 1974. From 1974 to 1981, as a member of the Horizontal Network Branch, he participated in the new adjustment of the North American Datum of 1983. His present position is Geodesist, Vertical Network Branch, and Project Manager, New Adjustment of the North American Vertical Datum of 1988.

Mr. Zilkoski is a member of the American Congress on Surveying and Mapping (ACSM) and is an instructor for the NGS Vertical Control Workshop. He is also a member of the American Geophysical Union and President of International Association of Geodesy Special Study Group 1.102, "Vertical Reference Systems."

ABSTRACT

Differential leveling observations are relative height differences measured between bench marks. Periodic adjustments are necessary as leveling observations are incorporated into the existing network. When combining the same order, class, and age of leveling data in relatively small network adjustments, the relative weighting scheme is usually not considered to be significant. In larger vertical control network adjustments, where many different orders and classes, as well as different ages of leveling data are combined, the relative weighting scheme is extremely important. If the a priori estimates of standard errors of leveling observations are incorrect, the observations will not receive the appropriate corrections and the adjusted heights will be incorrectly estimated. Also, post-adjustment error analysis can produce incorrect uncertainty values for adjusted results. An analysis of estimates of standard errors of leveling lines indicates that leveling data obtained by the National Geodetic Survey after 1978 are significantly more precise than data obtained in 1978 and earlier.

INTRODUCTION

Differential leveling observations are relative height differences measured between bench marks. The bench marks' heights and leveling observations are related through the following linear model:

$$H_j - H_i = L_b$$

where

H_i = height of bench mark i,

H_j = height of bench mark j, and

L_b = observed height difference between bench mark i and bench mark j.

In the North American Vertical Datum of 1988 (NAVD 88) readjustment project, as in most leveling networks, there are more observations than unknowns, i.e., the number of observed height differences exceeds the number of unknown bench mark heights. This redundancy determines the degrees of freedom of an adjustment, i.e., the degrees of freedom equals the number of observations minus the number of unknown parameters. Hence, different adjusted values of bench mark heights can be obtained by using different combinations of leveling data.

For the NAVD 88 project, the classical least squares method of observation equations is being be used to perform the adjustment of the leveling data. The mathematical model of the method of observation equations can be represented by the following equation:

$$L_a = F(X_a),$$

where L_a is a set of adjusted observations (e.g., leveling height differences), X_a is a set of parameters (e.g., bench mark heights), and F is a function which relates the observations to the parameters.

The set of observation equations can be represented as

$$V + L = AX,$$

where

V = vector of residuals (discrepancies),
A = design matrix,
X = vector of parameters, and
L = vector of observations.

It can be shown that when the least squares condition of minimum sum of the weighted residuals squared is fulfilled, the normal equations will be

$$NX + U = 0,$$

where

$N = A^T P A$,

$U = A^T P L$,

$P = k(\text{Var-Cov})^{-1}$,

Var-Cov = symmetric, positive definite, variance-covariance matrix of observations.

The least squares estimate of X is obtained from

$$X = -N^{-1}U.$$

In order to solve for X, the matrix N must be of full rank. In other words, the rank of N must be equal to the number of unknowns. In a leveling network adjustment consisting only of measured height differences, N will not be of full rank, but will actually be equal to the number of unknowns minus one. Therefore, at least one parameter will have to be weighted when using the method of weighted parameters. By weighting one parameter, fixing it to its a priori estimate, N can be inverted and the solution of X obtained. This is called a minimum constraint least squares adjustment, or "free" adjustment.

The observation equations for differential leveling observations between station i and station j consist of the following:

$$V_k = H_j - H_i - L_{ij},$$

where

V_k = residual for observation k,

H_i = height of bench mark i,

H_j = height of bench mark j, and

L_{ij} = observed height difference from station i to station j.

The observed height differences are assumed to be uncorrelated; hence all off-diagonal terms of the variance-covariance matrix of the observations are equal to zero. The nonzero, diagonal terms are not easy to determine. When combining the same order, class, and age of leveling data in relatively small network adjustments, the relative weighting scheme is usually not considered to be significant. In larger vertical control network adjustments, where many different orders and classes, as well as different ages of leveling data are combined, having a correct relative weighting scheme is extremely important. The a priori estimates of standard errors for all orders and classes of leveling data are assumed to be known. The weight of a leveling observation (p_i) is determined using the formula 1/(variance of the observation i), where the variance of the observation i is equal to: the a priori standard error squared times the distance leveled, in kilometers, divided by the number of runnings.

If the a priori estimates of standard errors of the leveling data are incorrect, the observations will not receive their appropriate corrections and the adjusted heights will be incorrectly estimated. In addition, observations may be incorrectly flagged as data outliers and removed from the analysis. The following basic assumptions are made when performing least squares adjustments:

(1) All data outliers have been removed from the data.
(2) The mathematical model is correct.
(3) Correct relative and absolute weights have been imposed.

All systematic errors must be resolved when evaluating the mathematical model. If one or more of these assumptions are not valid, the heights obtained from the adjustment may be distorted.

ESTIMATION OF STANDARD ERRORS

The a priori standard errors of 1 km of single-run leveling for first- and second-order leveling used by the National Geodetic Survey (NGS) are listed below:

first-order, class 0   = 0.7 mm,
first-order, class I   = 1.1 mm,
first-order, class II = 1.4 mm,

second-order, class I = 2.1 mm,
second-order, class II = 2.8 mm, and
second-order, class 0 = 3.0 mm.

The estimates of standard errors listed above were empirically determined in the late 1970s using a limited amount of data which were available in computer-readable form at the time the analysis was performed.

NGS' archival leveling data were processed and loaded into the NGS Integrated Data Base (NGSIDB) during the early 1980s. There are approximately 17,000 leveling lines in NGSIDB. In preparation for NAVD 88, a "standard error of 1 kilometer of single-run leveling" statistic was computed for each leveling line and loaded into NGSIDB. Approximately 14,000 leveling lines were used in this study. The 3,000 leveling lines not used consisted of Canadian leveling lines, and U.S. third-order data and single-run and second-order, class 0 area work which did not contain any double-run sections.

The formula given below was used to compute the standard error statistic:

$$\sigma_s = \left\{ \frac{1}{n} \sum_{i=1}^{n} \left[\frac{\sum_{j=1}^{m} [|\overline{\Delta h_i}| - |\Delta h_{ij}|]^2}{S_i(m-1)} \right] \right\}^{1/2}$$

where

σ_s = standard error of one km of single-run leveling,

n = number of sections with 2 or more non-rejected runnings,

m = number of non-rejected runnings in a section,

$\overline{\Delta h_i}$ = mean elevation difference for the ith section,

Δh_{ij} = the jth non-rejected elevation difference for section i, and

S_i = the length of section i in kilometers.

The leveling data were divided into six groups defined by major changes in leveling procedures and/or equipment as indicated below:

(1) Before 1902 - Use of "Vienna" or "Stampfer" type instruments and "paraffin" soaked wooden rods.

(2) Between 1902 and 1916 - New leveling instrument (Fischer level).

(3) Between 1917 and 1962 - New leveling rod (Invar rod).

(4) Between 1963 and 1970 - New type of instrument ("parallel-plate micrometer" instruments), new type of leveling rod (Kern double-scale Invar rods), and modified leveling procedures (reduced sight lengths and maximum differences in length of forward and backward sights at each setup were reduced).

(5) Between 1971 and 1978 - New leveling instrument (Zeiss Ni1 compensator instrument).

(6) After 1978 - New leveling instrument (Ni002 reversible compensator instrument), modified procedures (double-simultaneous, single-run leveling), introduction of automated recording system and low- and high-scale checks, and use of "motorized leveling" system.

DISCUSSION OF STANDARD ERROR ESTIMATES

The standard errors were plotted against the year the leveling lines was observed. Figure 1 depicts the standard error versus the year observed for first-order, class II leveling data. Figure 1 shows the obvious improvement in standard error estimates after 1970 for first-order, class II leveling data.

The weight of each line used in the weighted mean standard error estimate was computed using the formula: (number of runnings minus number of sections)/(number of sections). Thus, a leveling line that is double-run would get a weight of one. Table 1 gives the mean standard error estimate and weighted mean standard error estimate for all order and classes within each group for data in NGSIDB. Table 1 indicates that there is an improvement in standard errors for all data after 1970. For example, the standard error of first-order, class II leveling data for group 4 is 1.90, for group 5 it is 1.26, and for group 6 it is 1.01. This is a significant improvement in precision of leveling data. Similarly, the standard error of second-order, class I leveling data for group 5 is 1.28 and for group 6 it is 1.04. This implies that new procedures and/or instrumentation have improved the estimated precision of leveling data.

It should also be noted that the standard error estimates for second-order, class I leveling data are very similar to first-order, class II estimates for groups 5 and 6. For example, the standard error estimate for group 5 is 1.26 and the estimate for second-order, class I for group 5 is 1.28. This is probably because the specifications and procedures for performing first-order, class II and second-order, class I leveling are very similar. The only real differences in the procedures and specifications between the two orders and classes are the section and loop misclosure tolerances. Therefore, it is not surprising to find that the estimates of the standard errors for these two orders and classes are the same when a large sample of data is used.

In table 1, the statistics computed using all data are larger than the statistics computed using data which were obtained after 1978. What is also interesting to note is that the estimate for second-order, class I for all years is significantly less than the first-order, class II estimate for all years. This is because the standard errors are so large for first-order, class II leveling for groups 3 and 4 which contain over three-quarters of the total data for first-order, class II. This is a reason why the mean standard error estimated using all data should not be used for a priori estimates of standard errors. For example, the standard error estimated for first-order, class II leveling data using all years is 1.76 while the estimate is only 1.01 using data obtained after 1978.

The spread of the estimates of standard errors for first-order, class II appears to be large. The standard errors range from almost zero to 8.5. (See Figure 1). Figure 2 is a plot of the standard errors for first-order, class II leveling lines that were at least 50 percent double-run, i.e., the weight is equal to or greater than 0.5, against the number of bench marks on the leveling line. Notice that the large spread of standard errors appears to be for leveling lines containing less than 50 bench marks. This may indicate that for longer leveling lines, the computed statistic may be averaging out larger errors.

Table 2 gives the mean standard error estimates and weighted mean standard error estimate for all order and classes for each group using leveling lines that contain at least 50 bench marks and were at least 50 percent double-run. The standard errors increase in almost all cases, except for first-order, class I and first-order, class II leveling data. Although it should be noted that the sample size decreases significantly in all order and classes except for those two orders and classes that the standard errors values did not increase. Once again, this may indicate that the standard error statistic computed for longer leveling lines may be averaging out larger errors than the statistic implies. In another computation, the standard errors were also plotted against latitude and longitude to examine if the standard errors were location dependent. These plots did not show any apparent correlations.

CONCLUSION

Leveling data in NGS' Integrated Data Base that were obtained after 1978 appear to be significantly more precise than prior data. In addition, there tends to be an improvement in precision of all leveling data after an equipment and/or procedural change was imposed, indicating that the changes improved the precision of leveling data. The results of this study will be used to determine the standard errors of leveling data used in the NAVD 88. Special adjustments will be performed using preliminary standard errors values to assist in evaluating the results.

* * *

Mention of a commercial company or product does not constitute an endorsement by the National Oceanic and Atmospheric Administration. Use for publicity or advertising purposes of information from this publication concerning proprietary products or the tests of such products is not authorized.

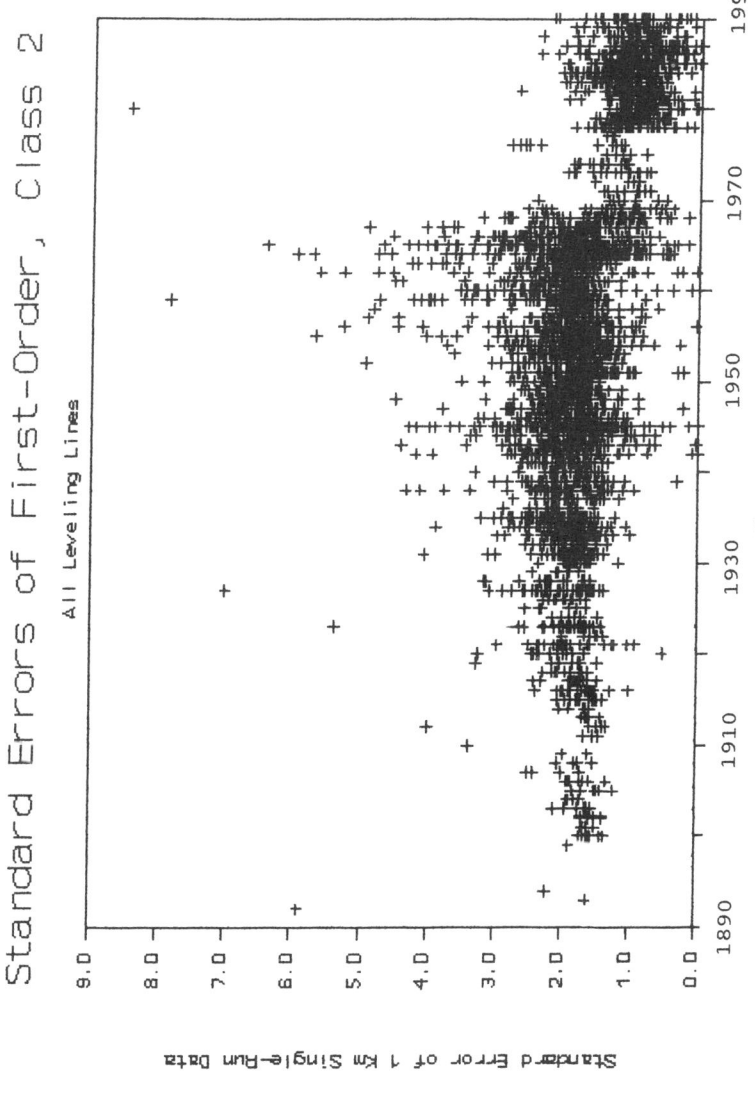

Figure 1. Plot of first-order, class II standard errors versus year observed

Table 1. Estimates of standard errors of leveling data in NGSIDB; all leveling lines were used to compute the statistics

Group	Order/Class					
	1/0	1/1	1/2	2/1	2/2	2/0
All Years	0.63 0.63 77	1.12 1.16 655	1.76 1.93 4,323	1.11 1.16 709	1.74 2.00 334	2.42 2.57 7,959
Before 1902	-- -- --	-- -- --	2.05 1.99 11	-- -- --	-- -- --	-- -- --
Between 1902 and 1916	-- -- --	-- -- --	1.73 1.78 98	-- -- --	-- -- --	1.94 1.94 1
Between 1917 and 1962	-- -- --	1.93 1.92 21	1.98 2.02 2613	1.16 1.75 8	2.24 2.24 1	2.42 2.58 6632
Between 1963 and 1970	0.83 0.86 10	1.19 1.23 250	1.90 1.95 668	2.14 2.14 1	2.57 2.59 2	2.47 2.62 1100
Between 1971 and 1978	0.83 0.83 31	1.06 1.08 329	1.26 1.39 109	1.28 1.33 206	1.77 2.10 192	1.97 2.17 226
After 1978	0.41 0.41 36	0.84 0.83 55	1.01 1.03 824	1.04 1.07 494	1.69 1.90 139	-- -- --

Weight - (No. Runs - No. Sections)/(No. Sections)
Group - Leveling lines which fall into age group were used in estimating the statistics.
1/0 - First-Order, Class 0
1/1 - First-Order, Class I
1/2 - First-Order, Class II
2/1 - Second-Order, Class I
2/2 - Second-Order, Class II
2/0 - Second-Order, Class 0

x.xx	- Mean standard error estimate
y.yy	- Weighted mean standard error estimate
nn	- Number of leveling lines used in estimating statistic.

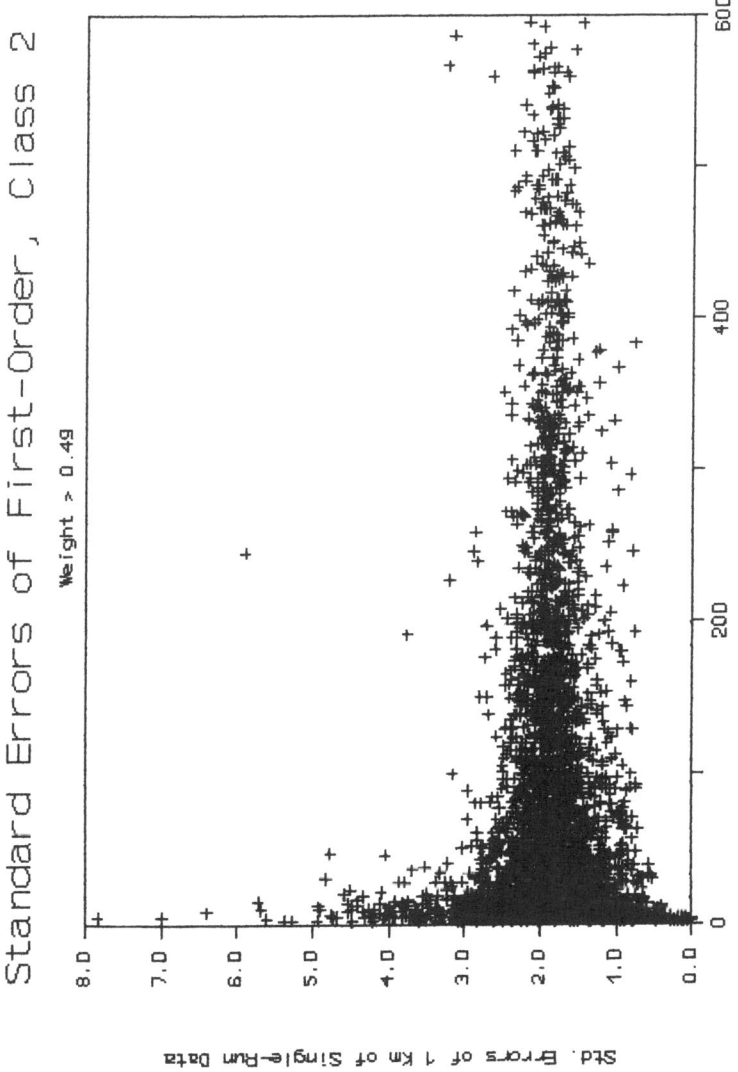

Figure 2. Plot of first-order, class II standard errors versus number of bench marks on leveling line

Table 2. Estimates of standard errors of leveling data in NGSIDB; only lines with at least 50 bench marks and at least 50 percent double-run were used to compute the statistics.

Group	Order/Class					
Lines with at least 50 BMs and at least 50% double-run	1/0	1/1	1/2	2/1	2/2	2/0
All Years	0.91	1.12	1.88	1.57	2.24	2.57
	0.92	1.14	1.90	1.62	2.29	2.56
	20	367	1,700	77	33	405
Before 1902	--	--	2.05	--	--	--
	--	--	1.99	--	--	--
	--	--	11	--	--	--
Between 1902 and 1916	--	--	1.68	--	--	1.94
	--	--	1.69	--	--	1.94
	--	--	79	--	--	1
Between 1917 and 1962	--	1.82	1.96	1.43	2.24	2.59
	--	1.83	1.98	1.43	2.24	2.59
	--	13	1,292	1	1	280
Between 1963 and 1970	1.13	1.16	1.75	--	--	2.54
	1.13	1.18	1.77	--	--	2.53
	2	130	231	--	--	92
Between 1971 and 1978	0.91	1.06	1.16	1.61	2.34	2.55
	0.90	1.07	1.17	1.63	2.40	2.51
	16	211	33	35	18	32
After 1978	0.73	0.92	1.06	1.54	2.10	--
	0.73	0.93	1.06	1.61	2.15	--
	2	13	54	41	14	--

Weight - (No. Runs - No. Sections)/(No. Sections)
Group - Leveling lines which fall into age group were used in estimating the statistics.
1/0 - First-Order, Class 0
1/1 - First-Order, Class I
1/2 - First-Order, Class II
2/1 - Second-Order, Class I
2/2 - Second-Order, Class II
2/0 - Second-Order, Class 0

x.xx - Mean standard error estimate
y.yy - Weighted mean standard error estimate
nn - Number of leveling lines used in estimating statistic.